# AJ Sadler

# Mathematics Methods

**1**

## Student Book

## Unit 1

NELSON
A Cengage Company

Australia • Brazil • Japan • Korea • Mexico • Singapore • Spain • United Kingdom • United States

Mathematics Methods Unit 1
1st revised Edition
A.J. Sadler

Publishing editor: Robert Yen
Project editor: Alan Stewart
Cover design: Chris Starr (MakeWork)
Text designers: Sarah Anderson, Nicole Melbourne,
Danielle Maccarone
Permissions researcher: Flora Smith
Answer checker: George Dimitriadis
Production controller: Erin Dowling
Typeset by: Cenveo Publisher Services
Reprint: Alice Kane

Any URLs contained in this publication were checked for currency
during the production process. Note, however, that the publisher
cannot vouch for the ongoing currency of URLs.

For product information and technology assistance,
in Australia call **1300 790 853**;
in New Zealand call **0800 449 725**

For permission to use material from this text or product, please email
**aust.permissions@cengage.com**

National Library of Australia Cataloguing-in-Publication Data
Sadler, A.J., author.
Mathematics methods : unit 1 / A.J Sadler.

1st revised edition
9780170390330 (paperback)
Includes index.
For secondary school age.

Mathematics--Study and teaching (Secondary)--Australia.
Mathematics--Textbooks.

510.712

Cengage Learning Australia
Level 7, 80 Dorcas Street
South Melbourne, Victoria Australia 3205

Cengage Learning New Zealand
Unit 4B Rosedale Office Park
331 Rosedale Road, Albany, North Shore 0632, NZ

For learning solutions, visit **cengage.com.au**

Printed in China by 1010 Printing International Ltd.
10 11 24

# PREFACE

This text targets Unit One of the West Australian course *Mathematics Methods*, a course that is organised into four units altogether, units one and two for year eleven and units three and four for year twelve.

The West Australian course, *Mathematics Methods*, is based on the Australian Curriculum Senior Secondary course *Mathematical Methods*. At the time of writing there is very little difference between the content of Unit One of the two courses, so this text would also be suitable for anyone following Unit One of the Australian Curriculum course, *Mathematical Methods*.

The book contains text, examples and exercises containing many carefully graded questions. A student who studies the appropriate text and relevant examples should make good progress with the exercise that follows.

The book commences with a section entitled **Preliminary work** to give the reader an early reminder of some of the work from earlier years that it will be assumed readers are familiar with, or for which the brief outline included in the section may be sufficient to bring the understanding of the concept up to the necessary level.

As students progress through the book they will encounter questions involving this preliminary work in the **Miscellaneous Exercises** that feature at the end of each chapter. These miscellaneous exercises also include questions involving work from preceding chapters to encourage the continual revision needed throughout the unit.

Students should also find that the content in some chapters involves work encountered in previous years, thus allowing speedy progress through those sections.

Some chapters commence with a '**Situation**' or two for students to consider, either individually or as a group. In this way students are encouraged to think and discuss a situation, which they are able to tackle using their existing knowledge, but which acts as a fore-runner and stimulus for the ideas that follow. Students should be encouraged to discuss their solutions and answers to these situations and perhaps to present their method of solution to others. For this reason answers to these situations are generally not included in the book.

At times in this series of books I have found it appropriate to go a little outside the confines of the syllabus for the unit involved. In this regard readers will find in this text that the Pythagorean theorem is assumed and used, the distance between two points with known coordinates is covered and the trigonometric identity $\sin^2 A + \cos^2 A = 1$ is included. When considering $y = a \sin x$, $y = \sin bx$ and $y = \sin (x - c)$, I also consider the more general $y = a \sin [b(x - c)] + d$.

The sine and cosine rules are considered early in this text so that students studying *Mathematics Specialist* cover this work in good time for its use in that course.

Alan Sadler

ISBN 9780170390330

# CONTENTS

# IMPORTANT NOTE

This series of texts has been written based on my interpretation of the appropriate *Mathematics Methods* syllabus documents as they stand at the time of writing. It is likely that as time progresses some points of interpretation will become clarified and perhaps even some changes could be made to the original syllabus. I urge teachers of the *Mathematics Methods* course, and students following the course, to check with the appropriate curriculum authority to make themselves aware of the latest version of the syllabus current at the time they are studying the course.

## Acknowledgements

As with all of my previous books I am again indebted to my wife, Rosemary, for her assistance, encouragement and help at every stage.

To my three beautiful daughters, Rosalyn, Jennifer and Donelle, thank you for the continued understanding you show when I am "still doing sums" and for the love and belief you show.

Alan Sadler

# PRELIMINARY WORK

This book assumes that you are already familiar with a number of mathematical ideas from your mathematical studies in earlier years.

This section outlines the ideas which are of particular relevance to Unit One of the *Mathematics Methods* course and for which some familiarity will be assumed, or for which the brief explanation given here may be sufficient to bring your understanding of the concept up to the necessary level.

Read this 'preliminary work' section and if anything is not familiar to you, and you don't understand the brief explanation given here, you may need to do some further reading to bring your understanding of those concepts up to an appropriate level for this unit. (If you do understand the work but feel somewhat 'rusty' with regards to applying the ideas some of the chapters afford further opportunities for revision, as do some of the questions in the miscellaneous exercises at the end of chapters.)

- Chapters in this book will continue some of the topics from this preliminary work by building on the assumed familiarity with the work.

- The **miscellaneous exercises** that feature at the end of each chapter may include questions requiring an understanding of the topics briefly explained here.

## Types of number

It is assumed that you are already familiar with counting numbers, whole numbers, integers, factors, multiples, prime numbers, composite numbers, square numbers, negative numbers, fractions, decimals, the rule of order, percentages, rounding to particular numbers of decimal places, truncating, the square root and the cube root of a number, powers of numbers (including zero and negative powers), and can use this familiarity appropriately. An ability to simplify simple expressions involving square roots is also assumed.

e.g. $\sqrt{8} = \sqrt{4 \times 2}$
$= 2\sqrt{2}$

$\sqrt{27} + \sqrt{75} = \sqrt{9 \times 3} + \sqrt{25 \times 3}$
$= 3\sqrt{3} + 5\sqrt{3}$
$= 8\sqrt{3}$

$\dfrac{3}{\sqrt{2}} = \dfrac{3}{\sqrt{2}} \times \dfrac{\sqrt{2}}{\sqrt{2}}$
$= \dfrac{3\sqrt{2}}{2}$

An understanding of numbers expressed in *standard form* or *scientific notation*, e.g. writing 260 000 in the form $2.6 \times 10^5$ or writing 0.0015 in the form $1.5 \times 10^{-3}$, is also assumed.

The set of numbers that you are currently familiar with is called the set of **real numbers**. We use the symbol $\mathbb{R}$ (or **R**) for this set. Sets like the whole numbers, the integers, the primes, the square numbers etc. are each subsets of $\mathbb{R}$.

# Direct proportion

If one copy of a particular book weighs 1.5 kg we would expect two copies of this book to weigh 3 kg, three to weigh 4.5 kg, four to weigh 6 kg, five to weigh 7.5 kg and so on. Every time we increase the number of books by one the weight goes up by 1.5 kg. Hence the straight line nature of the graph of this situation shown below.

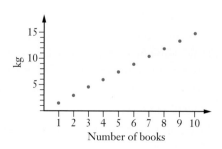

The number of books and the weight of the books are in **direct proportion** (also called **direct variation**):

> For two quantities that are in direct proportion, as one quantity is multiplied by a certain number then the other quantity is also multiplied by that number.
>
> - Doubling one will cause the other to double.
>
> - Halving one will cause the other to halve.
>
> - Trebling one will cause the other to treble. Etc.

# Inverse proportion

The graph below left shows the amount, $q$ kg, of a particular commodity that could be purchased for $1000 when the commodity costs $\$c$ per kg.

The graph below right shows the time taken, $t$ seconds, to travel a distance of 12 metres, by something travelling at $v$ metres/second.

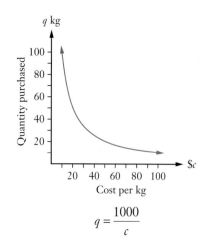

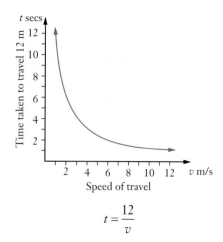

Rule:
$$q = \frac{1000}{c}$$

$$t = \frac{12}{v}$$

Each of the two previous situations involve **inverse proportion**:

> Double the cost of the commodity $\rightarrow$ Divide the quantity purchased by 2.
> Treble the cost of the commodity $\rightarrow$ Divide the quantity purchased by 3.     Etc.

> Double the speed of travel $\rightarrow$ Divide the time taken by 2.
> Treble the speed of travel $\rightarrow$ Divide the time taken by 3.     Etc.

If two variables are inversely proportional to each other the graph of the relationship will be that of a **reciprocal function**.

If $x$ and $y$ are inversely proportional the relationship will have an equation of the form

$$y = \frac{k}{x}.$$

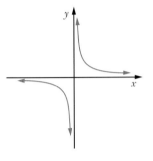

The graph will have the characteristic shape of a reciprocal relationship, as shown on the right, though in many applications negative values for the variables may not make sense and only that part of the typical shape for which both variables are positive will apply, as in the two earlier examples.

# Use of algebra

It is assumed that you are already familiar with manipulating algebraic expressions, in particular:

- **Expanding and simplifying**:

  For example     $4(x + 3) - 3(x + 2)$     expands to     $4x + 12 - 3x - 6$
      which simplifies to     $x + 6$

  $(x - 7)(x + 1)$     expands to     $x^2 + 1x - 7x - 7$
      which simplifies to     $x^2 - 6x - 7$

- **Factorising**:

  For example,     $21x + 7$     factorises to     $7(3x + 1)$
  $15apy + 12pyz - 6apq$     factorises to     $3p(5ay + 4yz - 2aq)$
  $x^2 - 6x - 7$     factorises to     $(x - 7)(x + 1)$
  $x^2 - 9$     factorises to     $(x - 3)(x + 3)$

  the last one being an example of the *difference of two squares* result:

  $x^2 - y^2$     factorises to     $(x - y)(x + y).$

You should also be familiar with the idea that solving an equation involves finding the value(s) the unknown can take that make the equation true.

For example, $x = 5.5$ is the solution to the equation     $15 - 2x = 4$
because     $15 - 2(5.5) = 4.$

Similarly, given two equations in two unknowns, for example: $\begin{cases} 5x - 2y = 6 \\ 3x + 2y = 26 \end{cases}$

'solving' means finding the pair of values that satisfy both equations, in this case the values are $x = 4$ and $y = 7$.

It is anticipated that you are familiar with solving various types of equation. For example:

**Linear equations**:

$3x - 5 = 7$       $6x + 1 = 15 - x$       $3(2x - 1) - (5 - 2x) = -20$

Solution:       $x = 4$       Solution:       $x = 2$       Solution:       $x = -1.5$

Equations with fractions:       $\dfrac{3x - 5}{2} = 8$       $\dfrac{21 - x}{2x + 3} = 2$

Solution:       $x = 7$       Solution:       $x = 3$

Simultaneous equations:       e.g.  $\begin{cases} y = 2x - 1 \\ y = x + 2 \end{cases}$       e.g.  $\begin{cases} 2x + y = 9 \\ 2x - 3y = 13 \end{cases}$

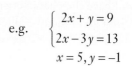

Solutions:       $x = 3, y = 5$       $x = 5, y = -1$

Simultaneous equations

**Quadratic equations** (including ones that are readily factorised):

e.g.       $x^2 = 25$       Solutions:       $x = \pm 5$

e.g.       $(2x - 3)(x + 1) = 0$       Solutions:       $x = 1.5, \quad x = -1$

e.g.       $x^2 - x - 20 = 0$

i.e.       $(x - 5)(x + 4) = 0$       Solutions:       $x = 5, \quad x = -4$

**Factorised cubics**:

e.g.       $x(x + 1)(x - 5) = 0$       Solutions:       $x = 0, \quad x = -1, \quad x = 5$

# Pascal's triangle

With an understanding of powers and an ability to expand brackets and to collect like terms we can show that:

$(x + y)^0 = 1$

$(x + y)^1 = x + y$

$(x + y)^2 = (x + y)(x + y) = x^2 + 2xy + y^2$

$(x + y)^3 = (x + y)(x + y)(x + y) = x^3 + 3x^2y + 3xy^2 + y^3$

$(x + y)^4 = (x + y)(x + y)(x + y)(x + y) = x^4 + 4x^3y + 6x^2y^2 + 4xy^3 + y^4$

$(x + y)^5 = (x + y)(x + y)(x + y)(x + y)(x + y) = x^5 + 5x^4y + 10x^3y^2 + 10x^2y^3 + 5xy^4 + y^5$

Noticing how the first six lines of **Pascal's triangle**, shown on the right, feature in the above expansions of $(x + y)^n$ for $n = 0, 1, 2, 3, 4$ and $5$, allows us, by using the appropriate lines of Pascal's triangle, to write down expansions of $(x + y)^n$ for higher values of $n$ directly. For example:

```
                           1
                        1     1
                     1     2     1
                  1     3     3     1
               1     4     6     4     1
            1     5    10    10     5     1
         1     6    15    20    15     6     1
      1     7    21    35    35    21     7     1
   1     8    28    56    70    56    28     8     1
1     9    36    84   126   126    84    36     9     1
```

$(x + y)^6 = x^6 + 6x^5y + 15x^4y^2 + 20x^3y^3 + 15x^2y^4 + 6xy^5 + y^6$

$(x + y)^8 = x^8 + 8x^7y + 28x^6y^2 + 56x^5y^3 + 70x^4y^4 + 56x^3y^5 + 28x^2y^6 + 8xy^7 + y^8$

# Function notation

Given the rule $y = 3x - 1$ and a particular value of $x$, say 5, we can determine the corresponding value of $y$, in this case 14.

The rule performs the function '*treble it and take one*' on any given $x$ value and outputs the corresponding $y$ value. It can be helpful to consider a function as a machine with a specific output for each given input:

Input  1, 2, 3, 4, 5, → The *treble it and take 1* function machine → Output 2, 5, 8, 11, 14,

In mathematics any rule that takes any given input value and assigns to it a particular output value is called a **function**.

We can write functions using the notation $f(x)$, pronounced '$f$ of $x$'.

For the 'treble it and take one' function we write $f(x) = 3x - 1$.

For this function:
$$f(1) = 3(1) - 1 = 2$$
$$f(2) = 3(2) - 1$$
$$= 5 \text{ etc.}$$

Alternatively we could use a second variable, say $y$, and express the rule as

$$y = 3x - 1.$$

The value of the variable $y$ depends on the value chosen for $x$. We call $y$ the **dependent variable** and $x$ the **independent variable**. The dependent variable is usually by itself on one side of the equation whilst the independent variable is 'wrapped up' in an expression on the other side.

# Types of function

From your mathematical studies of earlier years you should be familiar with:

## I Linear functions

These have:

- **equations** of the form $y = mx + c$.

  For example: $y = 3x - 1$ for which $m = 3$ and $c = -1$.

- **tables of values** for which each unit increase in the $x$ values sees a constant increase of $m$ in the corresponding $y$ values

  For example, for $y = 3x - 1$

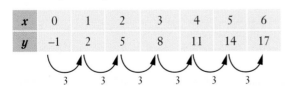

| $x$ | 0 | 1 | 2 | 3 | 4 | 5 | 6 |
|-----|----|----|----|----|----|----|----|
| $y$ | -1 | 2 | 5 | 8 | 11 | 14 | 17 |

- **graphs** that are straight lines with gradient $m$ and cutting the vertical axis at the point with coordinates $(0, c)$.

  For example, for $y = 3x - 1$

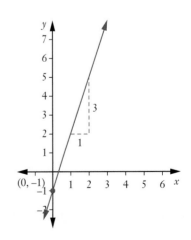

## II Quadratic functions

These have

- **equations** of the form $\qquad y = ax^2 + bx + c,\ a \neq 0$.

  For example, with $\qquad a = 1, b = 0$ and $c = 0$

  we have $\qquad y = x^2$ $\qquad$ the most basic quadratic.

  With $\qquad a = 2, b = -6$ and $c = 1$

  we have $\qquad y = 2x^2 - 6x + 1$.

  The equations of quadratic functions are sometimes written in the

  alternative forms $\qquad y = a(x - p)^2 + q$

  and $\qquad y = a(x - d)(x - e)$.

- **tables of values** with a constant *second difference* pattern

  For example, for $\qquad y = 2x^2 - 6x + 1$

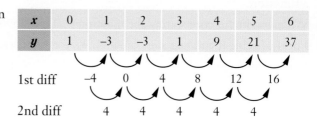

| $x$ | 0 | 1 | 2 | 3 | 4 | 5 | 6 |
|---|---|---|---|---|---|---|---|
| $y$ | 1 | −3 | −3 | 1 | 9 | 21 | 37 |

1st diff $\quad$ −4 $\quad$ 0 $\quad$ 4 $\quad$ 8 $\quad$ 12 $\quad$ 16

2nd diff $\quad$ 4 $\quad$ 4 $\quad$ 4 $\quad$ 4 $\quad$ 4

- **graphs** that are the same basic shape as that of $y = x^2$
  shown on the right, but that may be moved left, right,
  up, down, flipped over, squeezed or stretched.

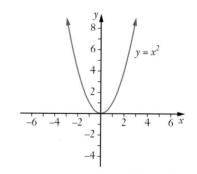

## III Reciprocal functions

- As mentioned a few pages earlier when we were considering inverse proportion, reciprocal functions have **equations** of the form

  $$y = \frac{k}{x}. \qquad \text{(Undefined for } x = 0\text{)}$$

  For example, with $k = 12$, $\qquad y = \frac{12}{x}$

- **tables of values** for which the $x$ and $y$ paired values have a common product (equal to $k$) for example, for $y = \frac{12}{x}$:

| $x$ | −4 | −3 | −2 | −1 | 0 | 1 | 2 | 3 | 4 |
|---|---|---|---|---|---|---|---|---|---|
| $y$ | −3 | −4 | −6 | −12 | Undefined | 12 | 6 | 4 | 3 |
| | −4 × −3 = 12 | −3 × −4 = 12 | −2 × −6 = 12 | −1 × −12 = 12 | | 1 × 12 = 12 | 2 × 6 = 12 | 3 × 4 = 12 | 4 × 3 = 12 |

ISBN 9780170390330

- **graphs** with the characteristic shape shown on the right, reflected in the $y$-axis if the $k$ in $y = \dfrac{k}{x}$ is negative.

# Pythagoras and trigonometry

It is anticipated that you have already encountered the Pythagorean theorem and the trigonometrical ratios of sine, cosine and tangent.

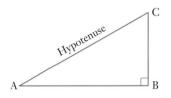

A very brief revision of the terminology and basic facts is included here.

In a right triangle we call the side opposite the right angle the **hypotenuse**.

The Pythagortean theorem states that:

*The square of the length of the hypotenuse of a right angled triangle is equal to the sum of the squares of the lengths of the other two sides.*

Thus, for the triangle shown, $AC^2 = AB^2 + BC^2$.

We refer to the other two sides of a right triangle as being **opposite** or **adjacent** to (next to) particular angles of the triangle.

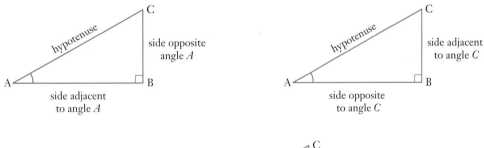

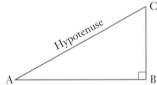

We then define the sine, cosine and tangent ratios as follows:

$$\sin A = \frac{\text{Opposite}}{\text{Hypotenuse}} = \frac{CB}{AC} \qquad\qquad \sin C = \frac{\text{Opposite}}{\text{Hypotenuse}} = \frac{AB}{AC}$$

$$\cos A = \frac{\text{Adjacent}}{\text{Hypotenuse}} = \frac{AB}{AC} \qquad\qquad \cos C = \frac{\text{Adjacent}}{\text{Hypotenuse}} = \frac{CB}{AC}$$

$$\tan A = \frac{\text{Opposite}}{\text{Adjacent}} = \frac{CB}{AB} \qquad\qquad \tan C = \frac{\text{Opposite}}{\text{Adjacent}} = \frac{AB}{CB}$$

The sine, cosine and tangent ratios can be remembered using the mnemonic **SOHCAHTOA**:

| SOH | CAH | TOA |
|---|---|---|
| $\text{Sin} = \dfrac{\textbf{O}\text{pposite}}{\textbf{H}\text{ypotenuse}}$ | $\text{Cos} = \dfrac{\textbf{A}\text{djacent}}{\textbf{H}\text{ypotenuse}}$ | $\text{Tan} = \dfrac{\textbf{O}\text{pposite}}{\textbf{A}\text{djacent}}$ |

Notice that it then follows that

$$\frac{\sin x}{\cos x} = \frac{\text{Opposite}}{\text{Hypotenuse}} \div \frac{\text{Adjacent}}{\text{Hypotenuse}}$$

$$= \frac{\text{Opposite}}{\text{Adjacent}}$$

$$= \tan x$$

The trigonometrical ratios of sine, cosine and tangent and the theorem of Pythagoras allow us to determine the lengths of sides and sizes of angles of right triangles, given sufficient information.

For example, given the diagram on the right, $x$ and $y$ can be determined as shown below:

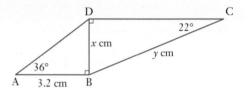

From $\triangle ABD$ $\qquad \tan 36° = \dfrac{x}{3.2}$

$\therefore \qquad\qquad\qquad x = 3.2 \tan 36°$

$\qquad\qquad\qquad\qquad \approx 2.32$

Correct to 1 decimal place, $x = 2.3$.

From $\triangle BCD$ $\qquad \sin 22° = \dfrac{x}{y}$

Hence $\qquad\qquad y \sin 22° = x$

and so $\qquad\qquad\qquad y = \dfrac{x}{\sin 22°}$

```
3.2 × tan36
                    2.32493609
Ans ÷ sin22
                    6.206340546
```

*Being sure to use the accurate value of $x$*, not the rounded value of 2.3, we obtain
$$y \approx 6.21.$$

Correct to 1 decimal place, $y = 6.2$.

With the usual convention for labelling a triangle, i.e. the angles use the capital letter of the vertex and lower case letters are used for sides opposite each angle, you may also be familiar with the following rules for $\triangle ABC$:

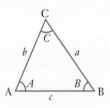

Area of a triangle: $\qquad \dfrac{ab \sin C}{2}$

The sine rule: $\qquad \dfrac{a}{\sin A} = \dfrac{b}{\sin B} = \dfrac{c}{\sin C}$

The cosine rule: $\qquad a^2 = b^2 + c^2 - 2bc \cos A$

These rules will be revised in this book in chapter 1, *Trigonometry*.

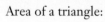

ISBN 9780170390330

An understanding of the use of **bearings** to indicate direction and of the concepts of an **angle of elevation** (from the horizontal, up) and **an angle of depression** (from the horizontal, down) is also assumed.

Thus in the diagram below left the angle of depression of point C from the top of the tower is 35° and in the diagram below right the angle of elevation of the top of the flagpole from point D is 15°.

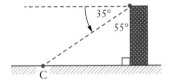 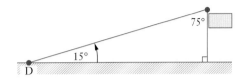

In the diagram on the right, from point A, the bearings of points B, C, D and E are shown in the table.

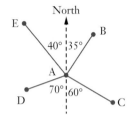

| Point | Three-figure bearing | Compass bearing |
|-------|---------------------|-----------------|
| B | 035° | N 35° E |
| C | 120° | S 60° E |
| D | 250° | S 70° W |
| E | 320° | N 40° W |

# Accuracy and trigonometry questions

Note that on the previous page the accurate value of *x* was used to determine *y*, not the rounded value, thus avoiding errors caused by premature rounding.

Note also that the answers for *x* and *y* were given as rounded values. Sometimes a situation may stipulate the degree to which answers must be rounded but if that is not the case you should round 'appropriately'. Just what is appropriate depends upon the accuracy of the data given and the situation. For example, in the calculation of the previous page it would be inappropriate to claim our value for *x* as 2.32493609, the answer obtained from a calculator, because it is far beyond the accuracy of the information used to obtain it, i.e. 3.2 cm and 36°.

In general our final answer should not be more accurate than the accuracy of the data we use to obtain it. The question gave us a length in cm, to 1 decimal place, so we should not claim greater accuracy for lengths we determine. Sometimes we may need to use our judgement of the likely accuracy of the given data. Given a length of 5 cm we might assume this has been measured to the nearest mm and hence give answers similarly to the nearest mm. (In theory a measurement of 5 cm measured to the nearest mm should be recorded as 5.0 cm but this is often not done.)

If the nature of the situation is known we might be able to judge the appropriate level of rounding. For example if asked to determine the dimensions of a metal plate that is to be made and then inserted into a patient, the level of accuracy may need to be greater than if dimensions were needed for some other situations.

In situations where accuracy is crucial any given measurements could be given with 'margins of error' included, for example 3.2 cm ± 0.05 cm, 36° ± 0.5°. More detailed error analysis could then be carried out and the margins of error for the answer calculated. However this is beyond the scope of this text.

# Sets

The **Venn diagram** on the right shows the **universal set**, U, which contains all of the **elements** currently under consideration, and the sets A and B contained within it.

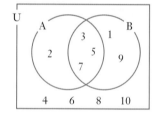

We use 'curly brackets' to list a set. Thus

$$U = \{1, 2, 3, 4, 5, 6, 7, 8, 9, 10\}$$

$$A = \{2, 3, 5, 7\} \text{ and } B = \{1, 3, 5, 7, 9\}$$

Set A has 4 members or elements.

We write $n(A) = 4$ or $|A| = 4$.

| | | |
|---|---|---|
| The number 9 is a member of B. | We write | $9 \in B$. |
| The number 2 is not in set B. | We write | $2 \notin B$. |
| $\{2, 7\}$ is a **subset** of A. | We write | $\{2, 7\} \subset A$. |

The order that we list the elements of a set is unimportant. The set {a, b, f } is the same set even if we list the three letters in a different order.

If a set has no elements it is said to be empty. We use ∅ as the symbol for an empty set. For example, {multiples of 4 that are odd numbers} = ∅.

If a set has an infinite number of members we say it is an infinite set. For example the positive integers form an infinite set, {1, 2, 3, 4, 5, 6, …}, as indicated by the '…'.

We use the symbol '∩' for the overlap or **intersection** of two sets.

Thus $A \cap B = \{3, 5, 7\}$

We use '∪' for the **union** of two sets.

Thus $A \cup B = \{1, 2, 3, 5, 7, 9\}$

We use A′ or $\overline{A}$ for the **complement** of A, i.e. everything in the universal set that is not in A. Thus $A' = \{1, 4, 6, 8, 9, 10\}$.

Venn diagrams can also provide a method for displaying information about the *number of elements* in various sets and can help to solve problems involving information of this kind. For example, suppose that 36 office workers were asked whether they had drunk tea or coffee during their morning breaks in the previous two weeks and further suppose that:

      7 said they had drunk both,

      8 said they had drunk only tea

and    6 said they had drunk neither.

This information could be used to create the Venn diagram shown on the right.

Asked how many had drunk coffee we can see from the Venn diagram that 22 had drunk coffee.

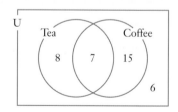

ISBN 9780170390330

In some questions of this type the information is not supplied in a 'nice' order and we may need to read on through the information before being able to accurately place a number in its space on the Venn diagram.

For example, in the previous tea/coffee situation suppose information had been presented as follows:

*A survey involving 36 office workers asked whether they had drunk tea or coffee during their morning breaks in the previous two weeks. The survey found that*

> *15 had drunk tea,*
>
> *22 had drunk coffee*

*and    6 said they had drunk neither.*

As we read the information we can 'note' it on the diagram, as shown by the bracketed numbers in the Venn diagram on the right. However we cannot accurately place numbers in the appropriate spaces until we reach the last piece of information, i.e. 6 said they had drunk neither.

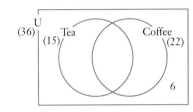

Only then can we complete the Venn diagram as shown.

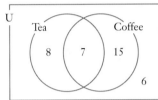

The Venn diagram on the right shows three sets A, B and C and the universal set U.

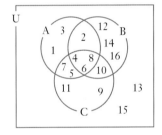

In this case

$$A = \{1, 2, 3, 4, 5, 6, 7, 8\}$$
$$A \cap B = \{2, 4, 6, 8\}$$
$$A \cup C = \{1, 2, 3, 4, 5, 6, 7, 8, 9, 10, 11\}$$
$$A \cap (B \cup C) = \{2, 4, 5, 6, 7, 8\}$$
$$A \cap B \cap C = \{4, 6, 8\}$$
$$A \cup B \cup C = \{1, 2, 3, 4, 5, 6, 7, 8, 9, 10, 11, 12, 14, 16\}$$

Why is it that we can write $A \cap B \cap C$ and not have to write this as $A \cap (B \cap C)$ or perhaps $(A \cap B) \cap C$? Why could we not do this with $A \cap (B \cup C)$?

Similarly why is it that we can write $A \cup B \cup C$ and not have to consider $A \cup (B \cup C)$ or $(A \cup B) \cup C$?

# Probability

Note:    The exercises in chapter 9, *Sets and probability*, contain questions that allow practice in some of the ideas about probability briefly explained here.

The probability of something happening is a measure of the likelihood of it happening and this measure is given as a number between zero (no chance of happening) to 1 (certain to happen).

In some cases we determine the probability of an event occurring by observing the outcome of a repeated number of trials in which the event is a possibility. The **long term relative frequency** with which the event occurs is then our best guess at the probability of the event occurring. Further trials may cause us to adjust this suggested probability.

For example if we flipped a biased coin five hundred times and found that it landed tail uppermost on 400 of these occasions we would suggest that in any one flip:

The probability that the coin lands with the tail uppermost $= \dfrac{400}{500} = \dfrac{4}{5}$.

This could also be expressed as a decimal, 0.8, or percentage, 80%.

Probability based on observed data like this is called **empirical probability**.

Activities such as rolling a die or flipping a coin are examples of **random phenomenon**. We are unable to consistently predict the outcome of a particular die roll or coin flip but when these activities are repeated a large number of times each has a predictable long run pattern.

The list of all possible outcomes that can occur when something is carried out is called the **sample space**. For example, for one roll of a normal fair die the sample space is: 1, 2, 3, 4, 5, 6.

The probability of an event occurring can be determined without the need for repeated experiment if we are able to present the sample space as a list of **equally likely outcomes**. We can then find a theoretical probability rather than an empirical probability.

Three common ways of presenting the equally likely outcomes are shown below.

| List | Table | Tree diagram |
|------|-------|--------------|
| Rolling a normal die once. | Roll two normal dice and sum the numbers. | Flip a coin three times and note outcome. |
| 6 equally likely outcomes. | 36 equally likely outcomes. | 8 equally likely outcomes. |
| 1, 2, 3, 4, 5, 6. | (table of dice sums) | (tree diagram of coin flips) |

Table of dice sums:

|   | 1 | 2 | 3 | 4 | 5 | 6 |
|---|---|---|---|---|---|---|
| 1 | 2 | 3 | 4 | 5 | 6 | 7 |
| 2 | 3 | 4 | 5 | 6 | 7 | 8 |
| 3 | 4 | 5 | 6 | 7 | 8 | 9 |
| 4 | 5 | 6 | 7 | 8 | 9 | 10 |
| 5 | 6 | 7 | 8 | 9 | 10 | 11 |
| 6 | 7 | 8 | 9 | 10 | 11 | 12 |

Tree diagram:

H — H — H → HHH
H — H — T → HHT
H — T — H → HTH
H — T — T → HTT
T — H — H → THH
T — H — T → THT
T — T — H → TTH
T — T — T → TTT

If we roll a normal die once the probability of getting a 3 is one-sixth. We write this as: $P(3) = \dfrac{1}{6}$.

An event occurring and it not occurring are said to be **complementary events**. If the probability of an event occurring is '$a$' then the probability of it not occurring is $1 - a$.

A Venn diagram may be used as a way of presenting the probability of events A and/or B occurring, as shown on the right.

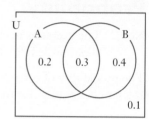

In this case:

$P(A) = 0.2 + 0.3$
$\quad = 0.5$

$P(B) = 0.3 + 0.4$
$\quad = 0.7$

$P(\bar{A}) = 0.4 + 0.1$
$\quad = 0.5$

$P(A \cup B) = 0.2 + 0.3 + 0.4$
$\quad = 0.9$

$P(A \cap B) = 0.3$

$P(U) = 0.2 + 0.3 + 0.4 + 0.1$
$\quad = 1$, as we would expect.

ISBN 9780170390330

# 1.

# Trigonometry

## Situation

A person sitting on a boat
- is situated 2 metres above sea level,
- has a device for measuring angles,

and
- notes that a straight line from themselves to the top most point of a nearby lighthouse makes an angle of 16° with the horizontal.

After travelling a further 50 metres directly away from the lighthouse this angle has decreased to 12°.

How high is the top most point of the lighthouse above sea level?

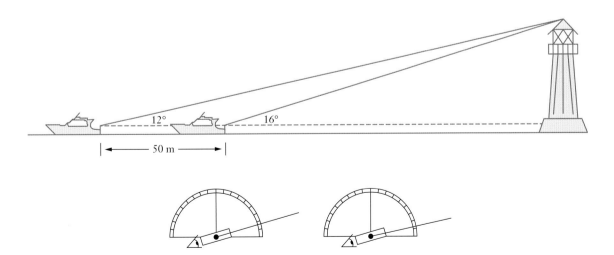

ISBN 9780170390330

iStock.com/Sergey Nepsha

How did you get on with the situation on the previous page?

Perhaps you applied trigonometry to the two right angled triangles and then solved the two equations this approach gave you.

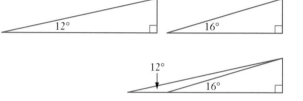

Perhaps you made a scale drawing and measured lengths from that.

Perhaps you have encountered the sine rule in earlier mathematical studies and applied that.

The *Preliminary work* stated that an ability to apply the trigonometrical ratios of sine, cosine and tangent to right angled triangles was assumed. In this chapter we will consider applying these ratios to triangles that are *not* right angled.

We will develop three formulae that can be applied to any triangles, right angled or not.

We will consider: • a formula for the area of a triangle,

        • the sine rule formula,

        • the cosine rule formula.

However we do have a problem:

Some non-right angled triangles will involve obtuse angles but from our right triangle definition of the sine of an angle, i.e.

$$\sin x° = \frac{\text{opposite}}{\text{hypotenuse}}$$

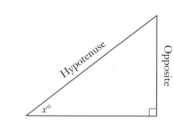

the idea of the sine of an angle bigger than 90° is meaningless because, in a right triangle, we cannot have angles bigger than 90°.

However, values for the sine and cosine of obtuse angles can be obtained from a calculator, as shown on the right.

Thus before we can use trigonometric ratios in triangles that might involve obtuse angles we need to consider what we mean by expressions like sin 95°, sin 150°, cos 100°, cos 170° etc.

| | |
|---|---|
| sin 95 | |
| | 0.9961946981 |
| sin 150 | |
| | 0.5 |
| cos 100 | |
| | −0.1736481777 |
| cos 170 | |
| | −0.984807753 |

ISBN 9780170390330

# The sine of an angle bigger than 90° – unit circle definition

The table below shows values for sin $x°$, as obtained from a calculator and rounded to two decimal places, for $x$ from 0 to 90 in steps of 10.

| $x$ | 0 | 10 | 20 | 30 | 40 | 50 | 60 | 70 | 80 | 90 |
|---|---|---|---|---|---|---|---|---|---|---|
| **sin $x°$ (2 dp)** | 0 | 0.17 | 0.34 | 0.50 | 0.64 | 0.77 | 0.87 | 0.94 | 0.98 | 1.00 |

If we are to redefine the sine of an angle to accommodate angles outside of 0° to 90° it makes sense to require that for angles between 0° and 90° any new definition gives the same values as our right triangle definition gives. Also we need any new definition to be useful, otherwise it simply will not 'stand the test of time'.

The definition of the sine and cosine of an angle that meets both of these requirements, i.e. it is consistent with the right triangle definition and it proves to be useful in its own right, is the **unit circle definition**.

Thew diagram on the right shows a circle centre O and of unit radius (i.e. a unit circle).

Point A is initially at location (1, 0) and the line OA, fixed at O, is rotated anticlockwise.

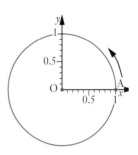

**The sine of the angle that AO makes with the positive $x$-axis is given by the $y$-coordinate of A.**

Thus for angles of 25°, 130° and 160° consider OA rotating to the positions shown in the diagrams below.

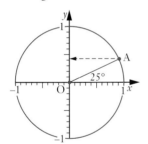 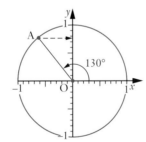 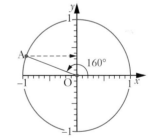

In each case the sine of the angle that AO makes with the positive $x$-axis is given by the $y$-coordinate of point A. Thus:

$$\sin 25° \approx 0.42 \qquad \sin 130° \approx 0.77 \qquad \sin 160° \approx 0.34.$$

Use the following diagrams to estimate values for sin 50°, sin 120° and sin 165° and then compare your answers to the values your calculator gives.

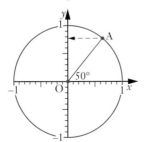 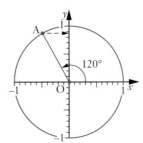 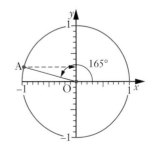

ISBN 9780170390330

The diagram on the right shows that for angles from 0° to 90° this unit circle definition for the sine of an angle gives exactly the same value as the right angled triangle definition.

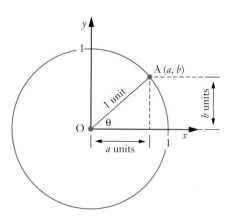

Note:  θ ('theta') is a Greek letter commonly used to represent angles.

**By the unit circle definition**

sin θ = the *y*-coordinate of A

$= b$

**By the right triangle definition**

$$\sin \theta = \frac{\text{opposite}}{\text{hypotenuse}}$$

$$= \frac{b}{1}$$

$$= b$$

Note also that from this unit circle definition, the sine of an obtuse angle is the same as the sine of its supplement. i.e. if A is an obtuse angle then

$$\sin A = \sin (180° - A)$$

For example

sin 155° = sin (180° − 155°)

= sin 25°

sin 130° = sin (180° − 130°)

= sin 50°

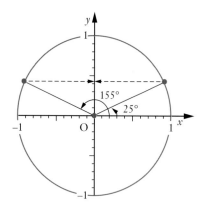

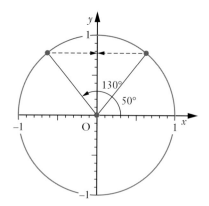

ISBN 9780170390330

# The cosine of an angle bigger than 90° – unit circle definition

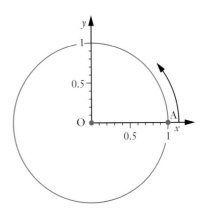

The diagram on the right shows a circle centre O and of unit radius.

Point $A$ is initially at location $(1, 0)$ and the line OA, fixed at O, is rotated anticlockwise.

**The cosine of the angle AO makes with the positive $x$-axis is given by the $x$-coordinate of A.**

Once again, for angles of 0° to 90° this unit circle definition for the cosine of an angle gives exactly the same value as the right angled triangle definition:

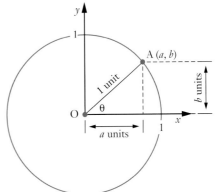

**By the unit circle definition**

$\cos \theta$ = the $x$-coordinate of A

$\quad = a$

**By the right triangle definition**

$$\cos \theta = \frac{\text{adjacent}}{\text{hypotenuse}}$$

$$= \frac{a}{1}$$

$$= a$$

From this unit circle definition we can see from the diagrams below that:

$\cos 35° \approx 0.82$  $\qquad\qquad$  $\cos 120° \approx -0.5$  $\qquad\qquad$  $\cos 145° \approx -0.82$

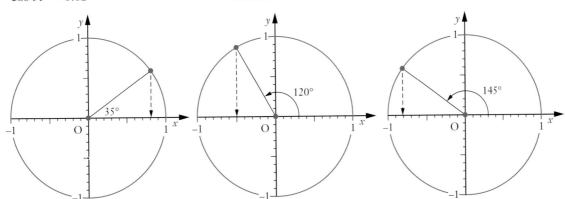

---

ISBN 9780170390330

**1.** Trigonometry ●●●●●●●●●●●

Note also that from this unit circle definition, the cosine of an obtuse angle is the negative of the cosine of its supplement. i.e. if A is an obtuse angle then

$$\cos A = -\cos(180° - A)$$

For example

$$\cos 155° = -\cos(180° - 155°)$$
$$= -\cos 25$$

$$\cos 130° = -\cos(180° - 130°)$$
$$= -\cos 50°$$

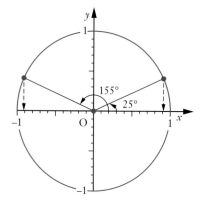
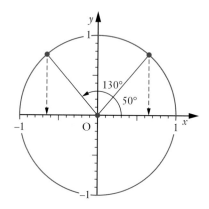

## Exercise 1A

Use the unit circle diagram shown to determine each of the following and, in each case, check the reasonableness of your answer by comparing it with the value given by your calculator.

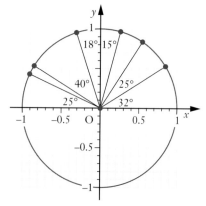

**1**  sin 0°          **2**  sin 32°          **3**  sin 57°

**4**  sin 75°          **5**  sin 90°          **6**  sin 108°

**7**  sin 148°          **8**  sin 155°          **9**  sin 180°

Use the unit circle diagram shown to determine each of the following and, in each case, check the reasonableness of your answer by comparing it with the value given by your calculator.

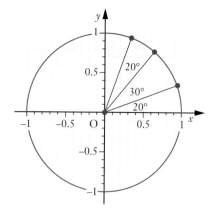

**10**  cos 0°          **11**  cos 20°          **12**  cos 50°

**13**  cos 70°          **14**  cos 90°          **15**  cos 160°

**16**  cos 180°          **17**  cos 110°          **18**  cos 130°

ISBN 9780170390330

# Area of a triangle given two sides and the angle between them

Suppose we have a triangle ABC and we know the lengths of two sides and the size of the angle between these sides, as shown below left for an acute angled triangle and below right for an obtuse angled triangle.

Acute angled triangle

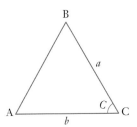

Draw the perpendicular from B
to meet AC at D.

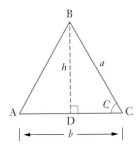

In $\triangle$BDC, $\quad \sin C = \dfrac{h}{a}$

Thus $\qquad h = a \sin C$

$\therefore \qquad$ Area $\triangle$ABC $= \dfrac{1}{2}$ base $\times$ height

$\qquad\qquad = \dfrac{1}{2} \times b \times a \sin C$

$\qquad\qquad = \dfrac{ab \sin C}{2}$

Obtuse angled triangle

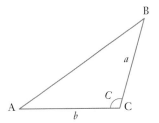

Draw the perpendicular from B
to meet AC produced at D.

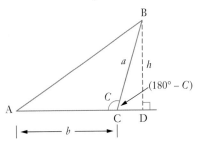

$\sin(180° - C) = \dfrac{h}{a}$

i.e. $\qquad \sin C = \dfrac{h}{a}$

Thus $\qquad\qquad h = a \sin C$

Hence: $\quad$ Area $\triangle$ABC $= \dfrac{1}{2} \times b \times a \sin C$

$\qquad\qquad\qquad = \dfrac{ab \sin C}{2}$

Thus for both acute and obtuse angled triangles:

$$\text{Area} = \frac{ab \sin C}{2}$$

i.e.: **The area of a triangle is half the product of two sides multiplied by the sine of the angle between them.**

ISBN 9780170390330

**1.** Trigonometry ●●○○○○○○○○

EXAMPLE 1

Find the area of the triangle shown.

**Solution**

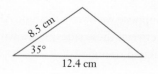

Area = $\dfrac{12.4 \times 8.5 \times \sin 35°}{2}$

$\approx 30.23$ cm$^2$

The area of the triangle is 30.2 cm$^2$, correct to one decimal place.

EXAMPLE 2

Find the area of the triangle shown.

**Solution**

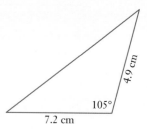

Area = $\dfrac{7.2 \times 4.9 \times \sin 105°}{2}$

$\approx 17.04$ cm$^2$

The area of the triangle is 17.0 cm$^2$, correct to one decimal place.

EXAMPLE 3

If each of the triangles shown below have an area of 7 cm$^2$ find $x$ correct to one decimal place in each case.

**a**

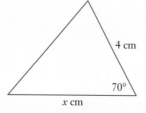

**b**

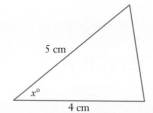

**Solution**

**a**     Area = $\dfrac{1}{2}(x)\,4\sin 70°$

$\therefore \quad 7 = \dfrac{1}{2}(x)\,4\sin 70°$

i.e.   $7 = 2x\sin 70°$

Solving this equation gives $x = 3.7$, correct to 1 decimal place.

**b**     Area = $\dfrac{1}{2}(4)(5)\sin x°$

$\therefore \quad 7 = \dfrac{1}{2}(4)(5)\sin x°$

i.e.   $7 = 10\sin x°$

Solving this equation gives $x = 44.4$ or $135.6$, correct to 1 decimal place.

ISBN 9780170390330

Note that for part **b** we had to give two answers because there are two possible values of $x$, in the range 0 to 180, for which $\sin x° = 0.7$. For one of these values $x°$ is an acute angle (44.4°) and for the other $x°$ is obtuse (180° − 44.4° = 135.6°).

Alternatively we could use the solve facility of some calculators to determine the answers to each part of the previous example, as shown on the right.

```
solve(7 = 2xsin(70), x)
                              {x = 3.724622204}

solve(7 = 10sin(x), x) | 0 ≤ x ≤ 180
          {x = 44.427004, x = 135.572996}
```

Note especially how both answers have been obtained for part **b** by instructing the calculator to solve for $x$ in the interval $0 \le x \le 180$.

The two solutions are shown below.

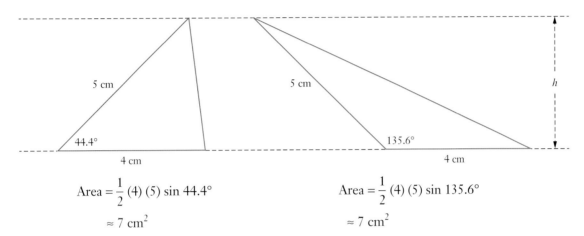

$$\text{Area} = \frac{1}{2}(4)(5)\sin 44.4°$$

$$\approx 7 \text{ cm}^2$$

$$\text{Area} = \frac{1}{2}(4)(5)\sin 135.6°$$

$$\approx 7 \text{ cm}^2$$

(Note that each triangle has the same length base, 4 cm, and the same height, $h$, and so must have identical areas, in this case 7 cm².)

When dealing with right triangles we knew that when needing to solve an equation like $\sin x° = 0.7$, the only applicable solution was the one for $x$ in the range 0 to 90.

With triangles that are not right angled we must be alert to the fact that an equation of the form $\sin x = c$ can have one solution for $x$ in the range 0 to 90 and another in the range 90 to 180.

## Exercise 1B

**1** Find $x$ in each of the following given that it is an acute angle. (Give your answer correct to the nearest degree.)

   **a**   $\sin x = 0.4$      **b**   $\sin x = 0.75$      **c**   $\sin x = 0.8$

**2** Find $x$ in each of the following given that it is an obtuse angle. (Give your answer correct to the nearest degree.)

   **a**   $\sin x = 0.2$      **b**   $\sin x = 0.3$      **c**   $\sin x = 0.55$

**3** Given that $x$ is between 0° and 180° state the two possible values $x$ can take. (Give your answer correct to the nearest degree.)

   **a**   $\sin x = 0.5$      **b**   $\sin x = 0.15$      **c**   $\sin x = 0.72$

Find the areas of each of the following triangles (not necessarily drawn to scale), giving your answers in square centimetres and correct to one decimal place.

**4**

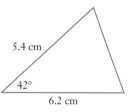

5.4 cm
42°
6.2 cm

**5**

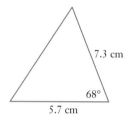

7.3 cm
68°
5.7 cm

**6**

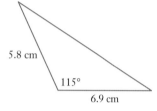

5.8 cm
115°
6.9 cm

**7**

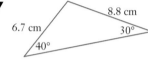

8.8 cm
6.7 cm
30°
40°

**8**

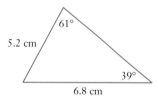

61°
5.2 cm
39°
6.8 cm

**9**

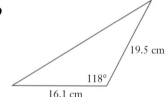

19.5 cm
118°
16.1 cm

Find the value of $x$, correct to one decimal place, in each of the following given that the area of each triangle is as stated. (The diagrams are not necessarily drawn to scale.)

**10**

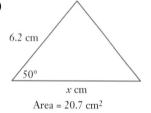

6.2 cm
50°
$x$ cm
Area = 20.7 cm$^2$

**11**

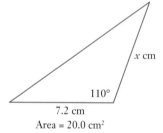

$x$ cm
110°
7.2 cm
Area = 20.0 cm$^2$

**12**

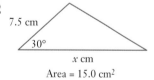

7.5 cm
30°
$x$ cm
Area = 15.0 cm$^2$

**13**

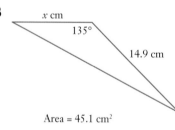

$x$ cm
135°
14.9 cm
Area = 45.1 cm$^2$

**14**

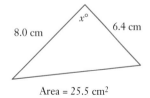

$x°$
8.0 cm
6.4 cm
Area = 25.5 cm$^2$

**15**

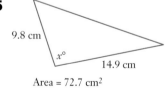

9.8 cm
$x°$
14.9 cm
Area = 72.7 cm$^2$

ISBN 9780170390330

# The sine rule

Consider a triangle ABC as shown below left for an acute angled triangle and below right for an obtuse angled triangle.

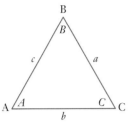

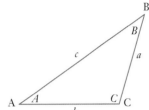

Drawing the perpendicular from B to meet AC at D.

Draw the perpendicular from $B$ to meet AC produced at D.

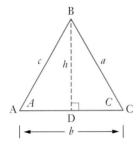

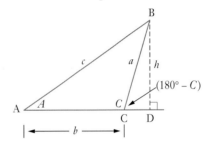

From $\triangle ABD$:     $\sin A = \dfrac{h}{c}$

$\therefore \qquad\qquad\qquad h = c \sin A \quad [1]$

From $\triangle CBD$:     $\sin C = \dfrac{h}{a}$

$\therefore \qquad\qquad\qquad h = a \sin C \quad [2]$

From $\triangle ABD$:     $\sin A = \dfrac{h}{c}$

$\therefore \qquad\qquad\qquad h = c \sin A \quad [1]$

From $\triangle CBD$:     $\sin (180° - C) = \dfrac{h}{a}$

$\therefore \qquad\qquad\qquad h = a \sin C \quad [2]$

Thus for both the acute triangle and the obtuse triangle:

From [1] and [2] $\qquad\qquad\qquad\qquad c \sin A = a \sin C$

Thus $\qquad\qquad\qquad\qquad\qquad \dfrac{c}{\sin C} = \dfrac{a}{\sin A} \qquad\qquad [3]$

If instead we draw the perpendicular from $A$ to $BC$ we obtain

$$\dfrac{b}{\sin B} = \dfrac{c}{\sin C} \qquad\qquad [4]$$

From [3] and [4] it follows that $\qquad \dfrac{a}{\sin A} = \dfrac{b}{\sin B} = \dfrac{c}{\sin C}$

This is **the sine rule**. Rather than learning this formula notice the pattern:

Any side on the sine of the opposite angle is equal to any other side on the sine of its opposite angle.

**EXAMPLE 4**

Find the value of $x$ in the following.

**a**

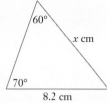

**b**

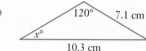

**Solution**

**a**  By the sine rule

$$\frac{x}{\sin 70°} = \frac{8.2}{\sin 60°}$$

Multiply by sin 70° to isolate $x$.

$$x = \frac{8.2 \sin 70°}{\sin 60°}$$

$$= 8.9 \text{ (to 1 decimal place)}$$

Or, using the 'solve' ability of some calculators:

$$\text{solve}\left(\frac{x}{\sin(70)} = \frac{8.2}{\sin(60)}, x\right)$$

$$\{x = 8.897521316\}$$

**b**  By the sine rule

$$\frac{10.3}{\sin 120°} = \frac{7.1}{\sin x°}$$

Multiply by $(\sin x°)(\sin 120°)$

$$10.3 \sin x° = 7.1 \sin 120°$$

$$\therefore \quad x \approx 36.7 \text{ (to 1 decimal place)}$$

Or, using the 'solve' ability of some calculators:

$$\text{solve}\left(\frac{10.3}{\sin(120)} = \frac{7.1}{\sin(x)}, x\right) \mid 0 \leq x \leq 180$$

$$\{x = 143.346877, x = 36.65312298\}$$

Note:

- In part **b** we say $x = 36.7$ despite there being another value of $x$ between 0 and 180 for which $10.3 \sin x° = 7.1 \sin 120°$, i.e. $x = 180 - 36.7$, as the calculator shows when asked for solutions in the interval $0 \leq x \leq 180$. However, in the given triangle, $x$ cannot be 143.3 because the triangle already has one obtuse angle and cannot have another. However we will not always be able to dismiss this other value as part (a) of the next example shows.

- Some calculator programs and internet websites allow the user to put in the known sides and angles of a triangle and, provided the information put in is sufficient, the program will determine the remaining sides and angles.

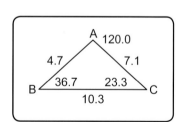

Some calculators allow us to create a scale drawing of the triangle and find lengths and angles that way. These programs can be useful and are worth exploring but make sure that you understand the underlying idea of the sine rule (and the cosine rule which we will see later in this chapter) and can demonstrate the appropriate use of these rules when required to do so.

**EXAMPLE 5**

Find the value of $x$ in the following.

**a**

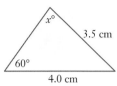

**b**

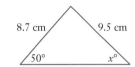

**Solution**

**a**  By the sine rule

$$\frac{4.0}{\sin x°} = \frac{3.5}{\sin 60°}$$

Multiply by $(\sin x°)(\sin 60°)$

$4.0 \sin 60° = 3.5 \sin x°$

$\therefore \quad \sin x° = \frac{4.0 \sin 60°}{3.5}$

$x \approx 81.8$ or $98.2$,

both of which are possible for the given information.

Or, using the 'solve' ability of some calculators:

$$\text{solve}\left(\frac{4}{\sin(x)} = \frac{3.5}{\sin(60)}, x\right) \Big| 0 \le x \le 180$$

$$\{x = 81.7867893, x = 98.2132107\}$$

**b**  (Note that $x°$, being opposite a side of length 8.7 cm, must be less than the 50° which is opposite a side of length 9.5 cm.)

$$\frac{9.5}{\sin 50°} = \frac{8.7}{\sin x°}$$

Multiply by $(\sin 50°)(\sin x°)$

$9.5 \sin x° = 8.7 \sin 50°$

$\therefore \quad \sin x° = \frac{8.7 \sin 50°}{9.5}$

Thus $\quad x \approx 44.6$

Or, using the 'solve' ability of some calculators:

$$\text{solve}\left(\frac{9.5}{\sin(50)} = \frac{8.7}{\sin(x)}, x\right) \Big| 0 \le x \le 180$$

$$\{x = 135.4496775, x = 44.55032253\}$$

Note that **Example 5** part **a** is similar to **Example 3** part **b** in that there are two triangles that fit the given information. We need to be alert to the possibility of this second solution when the sine rule leads to an equation that is of the form $\sin x = c$.

However note also that in part **b** we could dismiss the obtuse angled solution because $x$ had to be smaller than 50. (Or, had we not noticed this from the side lengths, we would reject the obtuse angle as the angle sum of the triangle would exceed 180°.)

Example 5 **a** is an example of the "ambiguous case" that can arise when using the sine rule. The two triangles that fit the given information are shown.

In $\triangle ABC_1$, angle $A = 60°$, $AB = 4.0$ cm and $BC_1 = 3.5$ cm.

In $\triangle ABC_2$, angle $A = 60°$, $AB = 4.0$ cm and $BC_2 = 3.5$ cm.

In $\triangle ABC_1$, angle $AC_1B \approx 98.2°$.

In $\triangle ABC_2$, angle $AC_2B \approx 81.8°$.

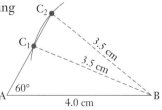

**1.** Trigonometry ●●●●●●●●●●

# The cosine rule

Again consider a triangle ABC as shown below left for an acute angled triangle and below right for an obtuse angled triangle.

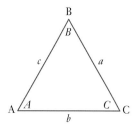

Again we draw the perpendicular from B to meet AC at D.

Again we draw the perpendicular from B to meet AC produced at D.

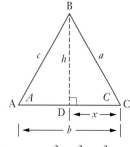

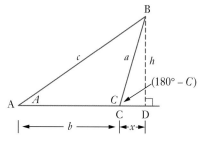

From $\triangle$CBD:  $\quad a^2 = h^2 + x^2$  [1]

From $\triangle$ABD:  $\quad c^2 = h^2 + (b - x)^2$

i.e.  $\quad c^2 = h^2 + b^2 + x^2 - 2bx$

Using [1]:  $\quad c^2 = a^2 + b^2 - 2bx$  [2]

From $\triangle$CBD:  $\quad \cos C = \dfrac{x}{a}$

$\therefore \quad x = a \cos C$  [3]

Using [2] and [3]:  $\quad c^2 = a^2 + b^2 - 2ab \cos C$

From $\triangle$CBD:  $\quad a^2 = h^2 + x^2$  [1]

From $\triangle$ABD:  $\quad c^2 = h^2 + (b + x)^2$

i.e.  $\quad c^2 = h^2 + b^2 + x^2 + 2bx$

Using [1]:  $\quad c^2 = a^2 + b^2 + 2bx$  [4]

From $\triangle$CBD:  $\quad \cos (180° - C) = \dfrac{x}{a}$

$\therefore \quad x = -a \cos C$  [5]

Using [4] and [5]:  $\quad c^2 = a^2 + b^2 - 2ab \cos C$

Thus for both the acute triangle and the obtuse triangle $c^2 = a^2 + b^2 - 2ab \cos C$.

This is **the cosine rule**.   $\quad c^2 = a^2 + b^2 - 2ab \cos C$

Similarly   $\quad a^2 = b^2 + c^2 - 2bc \cos A \quad$   and   $\quad b^2 = a^2 + c^2 - 2ac \cos B$

As was said with the sine rule, rather than learning the rule as a formula instead notice the pattern of what it is telling you:

The square of any side of a triangle is equal to the sum of the squares of the other two sides take away twice the product of the other two sides multiplied by the cosine of the angle between them.

ISBN 9780170390330

## EXAMPLE 6

Find the value of $x$ for the triangle shown sketched.

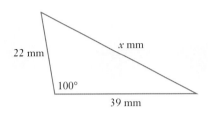

22 mm

$x$ mm

100°

39 mm

**Solution**

By the cosine rule:

$x^2 = 22^2 + 39^2 - 2 (22) (39) \cos 100°$

$\approx 2302.98$

$x = 48$ to the nearest integer.

$$22^2 + 39^2 - 2 \times 22 \times 39 \times \cos(100)$$
$$2302.980273$$
$$\sqrt{\text{ans}}$$
$$47.98937667$$

## EXAMPLE 7

Find the value of $x$ for the triangle shown sketched.

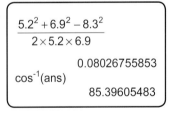

$x°$

5.2 cm

6.9 cm

8.3 cm

**Solution**

By the cosine rule:

$8.3^2 = 5.2^2 + 6.9^2 - 2 (5.2) (6.9) \cos x°$

$\cos x° = \dfrac{5.2^2 + 6.9^2 - 8.3^2}{2(5.2)(6.9)}$

$\approx 0.080\,27$

$x = 85$ to the nearest integer.

$$\dfrac{5.2^2 + 6.9^2 - 8.3^2}{2 \times 5.2 \times 6.9}$$
$$0.08026755853$$
$$\cos^{-1}(\text{ans})$$
$$85.39605483$$

Note:

- If you prefer to use the solve facility on your calculator make sure you can obtain the same answers as those shown above.

- In **Example 7** above there was no need to consider the possibility of a second solution arising, as we had to do when using the sine rule, because if the angle had been obtuse we would have had a negative value for its cosine. An equation of the form $\cos x = c$ does *not* have two solutions for $x$ in the range 0° to 180°.

EXAMPLE 8

The sketch on the right shows a system of three triangles with lengths and angles as indicated.

BAE is a straight line.
Find the length of CD.

**Solution**

**Thoughts**

CD is one side of △ACD. In this triangle we know the lengths of AC and AD so if we know the size of ∠CAD we could apply the cosine rule to find the length of CD. We can find the size of ∠CAD if we first find the size of ∠CAB and the size of ∠DAE.

For △ABC, applying the cosine rule:

$$68^2 = 41^2 + 37^2 - 2 \times 41 \times 37 \cos \angle BAC$$

$$\cos \angle BAC = \frac{41^2 + 37^2 - 68^2}{2 \times 41 \times 37}$$

∴   ∠BAC ≈ 121.3°

```
(41² + 37² − 68²) ÷ (2 × 41 × 37)
                           −0.5187870798
cos⁻¹Ans
                           121.2509263
Ans → A
                           121.2509263
```

For △DAE, applying the cosine rule:

$$21^2 = 33^2 + 48^2 - 2 \times 33 \times 48 \cos \angle DAE$$

$$\cos \angle DAE = \frac{33^2 + 48^2 - 21^2}{2 \times 33 \times 48}$$

∴   ∠DAE ≈ 21.3°

```
(33² + 48² − 21²) ÷ (2 × 33 × 48)
                           0.9318181818
cos⁻¹Ans
                           21.27996647
Ans → B
                           21.27996647
```

For △CAD, applying the cosine rule:

$$CD^2 = 33^2 + 41^2 - 2 \times 33 \times 41 \cos \angle CAD$$

$$\approx 622.3$$

∴   CD ≈ 24.9

```
33² + 41² − 2 × 33 × 41cos(180−A−B)
                           622.297976
√Ans
                           24.94590099
```

CD is of length 25 mm, to the nearest millimetre.

Notice from the calculator displays that the more accurate values for ∠BAC and ∠DAE were stored and later recalled for use, thus avoiding the risk of introducing unnecessary rounding errors.

**EXAMPLE 9**

Solve the triangle ABC given that AB is of length 6.2 cm, AC is of length 7.1 cm and angle $A = 35°$.

**Solution**

First make a sketch:

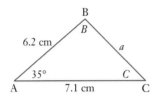

By the cosine rule
$$a^2 = 7.1^2 + 6.2^2 - 2\,(7.1)\,(6.2) \cos 35°$$

∴
$$a \approx 4.09 \text{ cm}$$

By the sine rule
$$\frac{a}{\sin 35°} = \frac{6.2}{\sin C}$$

∴ $C \approx 60.4°$ (Obtuse angle not applicable – see point below.)

Thus $B = 180° - (35° + C)$

$\approx 84.6°$

In △ABC, $\angle B = 85°$, $\angle C = 60°$ and $a = 4.1$ cm, all angles to the nearest degree and lengths correct to one decimal place.

**Important point**

In the last example, having used the cosine rule to determine the length of BC we next used the sine rule to find angle $C$ rather than angle $B$. This was because $C$, being opposite a smaller side, could not be obtuse. This allowed us to say with confidence that $C \approx 60.4°$ and we did not have to consider $(180 - 60.4)°$.

# Some vocabulary

If a question refers to a line **subtending** an angle at a point this is the angle formed by joining each end of the line to the point.

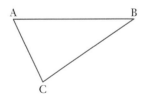

Line AB subtends $\angle ACB$ at C.

Chord DE subtends $\angle DFE$ at the centre, F.

If a set of points is referred to as being **collinear** this means they lie in a straight line.

If all four vertices of a quadrilateral lie on the circumference of a circle the quadrilateral is said to be a **cyclic quadrilateral**. One of the properties of cyclic quadrilaterals is that their opposite angles add up to 180°.

*Finding an unknown side*

*Finding an unknown angle*

## Exercise 1C

### The sine rule

Find the value(s) of $x$ in each of the following. (Diagrams not necessarily drawn to scale.)

**1**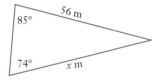

**2**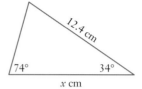

**3**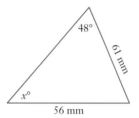

**4**

**5**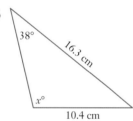

**6**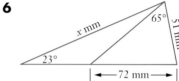

**7**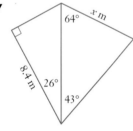

**8**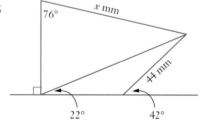

**9** The diagram shows a pole AB with end A fixed on horizontal ground and the pole supported by a wire attached to end B and to a point C on the ground with AC = 420 centimetres.

The pole makes an angle of 50° with the ground and the wire makes an angle of 30° with the ground, as shown in the diagram. The points A, B and C all lie in the same vertical plane.

Find the length of the pole giving your answer to the nearest centimetre.

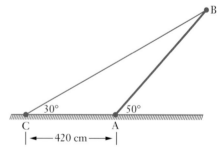

**10** Rather than risking the direct shot over a lake a golfer prefers to take two shots to get to the green as shown in the diagram on the right. How much further is this two shot route than the direct route?

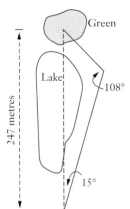

### The cosine rule

Find the value of $x$ in each of the following. (Diagrams not necessarily drawn to scale.)

When $x$ involves an angle give your answer to the nearest whole number.

When $x$ involves a length give an accuracy consistent with the given lengths.

**11**

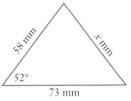

58 mm, $x$ mm, 52°, 73 mm

**12**

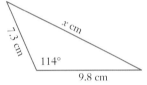

7.3 cm, $x$ cm, 114°, 9.8 cm

**13**

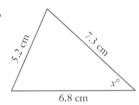

5.2 cm, 7.3 cm, $x°$, 6.8 cm

**14**

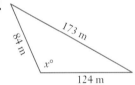

84 m, 173 m, $x°$, 124 m

**15**

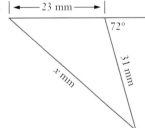

23 mm, 72°, $x$ mm, 31 mm

**16**

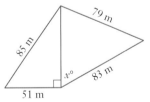

85 m, 79 m, $x°$, 83 m, 51 m

**17**

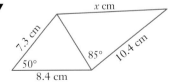

$x$ cm, 7.3 cm, 50°, 85°, 10.4 cm, 8.4 cm

**18**

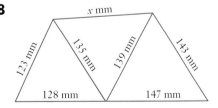

$x$ mm, 123 mm, 135 mm, 139 mm, 143 mm, 128 mm, 147 mm

**19** A boat travels 6.3 km due North and then turns 17° towards the West and travels a further 7.2 km. How far is it then from its initial position?

**20** Jim and Toni leave the same point at the same time with Jim walking away at a speed of 1.4 m/s and Toni at a speed of 1.7 m/s, the two directions of travel making an angle of 50° with each other. If they both continue on these straight line paths how far are they apart after 8 seconds?

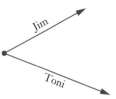

### Miscellaneous

Find the value of $x$ in each of the following. (Diagrams not necessarily drawn to scale.)

**21**

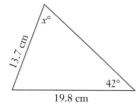

**22**

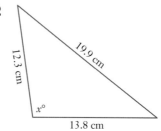

**23**

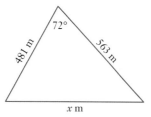

**24**

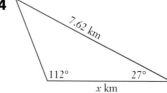

**25**

**26**

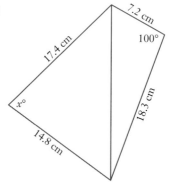

**27**

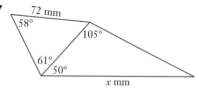

**28**

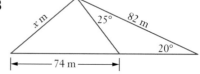

ISBN 9780170390330

**29** The diagram on the right shows a mobile crane used to lift containers from ships and transfer them to waiting container trucks. If AB is of length 300 centimetres find the lengths of AC and BC.

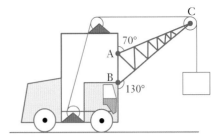

**30** A triangle has sides of length 12.7 cm, 11.9 cm and 17.8 cm. Find the size of the smallest angle of the triangle, giving your answer to the nearest degree.

**31** In $\triangle$ABC, $\angle A = 72°$, $b = 7.3$ cm and $a = 9.1$ cm. Find the length of AB.

**32** Solve $\triangle$ABC given that $\angle A = 43°$, $c = 12.4$ cm and $b = 14.3$ cm.

**33** From a lighthouse, ship A is 15.2 kilometres away on a bearing 030° and ship B is 12.1 kilometres away on a bearing 100°. How far is B from A?

**34** From a lighthouse, ship P is 7.3 km away on a bearing 070°. A second ship Q is on a bearing 150° from P and 130° from the lighthouse. How far is Q from the lighthouse?

**35** A tower stands vertically at the base of a hill that inclines upwards at 30° to the horizontal. From a point 25 metres from the base of the tower, and directly up the hill, the tower subtends an angle of 52°. Find the height of the tower giving your answer correct to the nearest metre.

**36** A parallelogram has sides of length 3.7 cm and 6.8 cm and the acute angle between the sides is 48°. Find the lengths of the diagonals of the parallelogram.

**37** The diagonals AC and BD, of parallelogram ABCD, intersect at E.

If $\angle$AED = 63° and the diagonals are of length 10.4 cm and 14.8 cm use the fact that the diagonals of a parallelogram bisect each other to determine the lengths of the sides of the parallelogram.

**38** The tray of the tip truck shown on the right is tipped by the motor driving rod BC clockwise about B. As the tray tips end C moves along the guide towards A.

If AB = 2 metres and BC = 1 metre find the size of $\angle$CAB when AC is

**a** 2.6 metres      **b** 2.1 metres

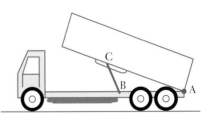

**39** The 'W-type roof truss' shown on the right is to be constructed with AE = 900 cm and AG = GF = FE.

∠DEF = 20°, ED = DC and the truss is symmetrical with the vertical line through $C$ as the line of symmetry.

Calculate the following lengths, correct to the nearest cm.

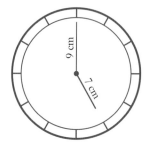

**a**   CE          **b**   ED          **c**   DF          **d**   CF

**40** Find, to the nearest millimetre, the distance between the tip of the 70 mm hour hand and the tip of the 90 mm minute hand of a clock at:

**a**   5 o'clock

**b**   10 minutes past 5

**41** A coastal observation position is known to be 2.50 km from a lighthouse. The coastguard in the observation position is in radio and visual contact with a ship in distress at sea. If the coastguard looks towards the lighthouse and then towards the ship these two directions make an angle of 40° with each other. If the captain on the ship looks towards the observation position and then towards the lighthouse these two directions make an angle of 115° with each other. (The ship, the lighthouse and the observation position may all be assumed to be on the same horizontal level.)

How far is the ship from:

**a**   the lighthouse?

**b**   the coastal observation position?

**42** Do the following question twice, once using Pythagoras and the cosine rule and once using right triangle trigonometry.

Find, to the nearest degree, the size of the largest of the three angles of △ABC where A, B and C have coordinates A(6, 2), B(2, 5), C(−6, −3).

**43** A, B and C are three collinear points on level ground with B between A and C. The distance from A to B is 40 m. A vertical tower, CD, has its base at C. From A and B the top of the tower, point D, has angles of elevation of 20° and 35° respectively. Find the height of the tower.

**44** Points A, B and C all lie on the same horizontal ground with B due north of A and C on a bearing of 030° from A.

From the top of a vertical tower at A, 37 metres above ground, point B has an angle of depression of 17° and point C has an angle of depression of 12°.

How far is B from C?

ISBN 9780170390330

**45** Solve $\triangle GHI$ given that $\angle G = 55°$, $i = 19.4$ cm and $g = 18.2$ cm.

**46** Points A, B and C lie on horizontal ground. From A the bearings of B and C are 330° and 018° respectively. A vertical tower of height 40 m has its base at A. From B and C the angles of elevation of the top of the tower are 20° and 12° respectively. How far is B from C, to the nearest metre?

**47** ABCD is a cyclic quadrilateral with $\angle DAB = 100°$, AB = 7.2 cm, AD = 6.1 cm and BC = 8.2 cm. Find:

   **a** the size of $\angle BCD$,

   **b** the size of $\angle ADC$,

   **c** the perimeter of the quadrilateral,

   **d** the area of the quadrilateral.

**48** ABCD is a cyclic quadrilateral with AB = 10 cm, BC = 12 cm, CD = 9 cm, DA = 14 cm. If $\angle ABC = \theta$, $\angle ADC = \phi$ and AC is of length $x$ cm, find:

   **a** an expression for $x^2$ in terms of $\cos \theta$,

   **b** an expression for $x^2$ in terms of $\cos \phi$,

   **c** $\theta$ in degrees correct to the nearest degree.

**49** Make use of the cosine rule, and the rule for the area of a triangle given two sides and the included angle, to determine the area of a triangular block of land with sides of length 63 m, 22 m and 55 m and then check that your answer agrees with the following statement (known as *Heron's rule*):

The area of a triangle with sides of length $a$, $b$ and $c$ is given by:

$$\text{Area} = \sqrt{s(s-a)(s-b)(s-c)} \text{ where } s = \frac{a+b+c}{2}.$$

**50** A triangular building block has sides of length 25 metres, 48 metres and 53 metres. A second triangular block has sides of length 33 metres, 38 metres and 45 metres. Which block has the greater area and by how much (to the nearest square metre)?

---

## Regular polygons

Suppose that a regular $n$-sided polygon has all of its vertices touching the circumference of a circle of radius 1 unit. For $n = 5, 6, 7$ and 8 this is shown below:

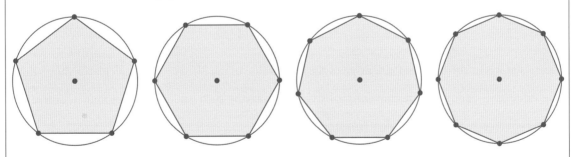

Find the area of each of the above polygons and investigate this situation for increasing integer values of $n$.

---

ISBN 9780170390330

# Exact values

It can sometimes be the case that a mathematician is asked to solve an equation and, not knowing for what purpose the answer is required, is not sure to what degree of accuracy the answer should be given. To avoid having to make this decision regarding accuracy the mathematician may choose to leave the expression as an **exact value**. Others can then use this exact value to whatever degree of accuracy they require.

For example, suppose we were asked to solve the equation:
Rather than deciding to give the answer to, say, 2 decimal places
we could instead give the answer in *exact* form:

$$x^2 - 1 = 5$$
$$x = \pm 2.45$$
$$x = \pm \sqrt{6}.$$

In trigonometry, commonly-used exact values are the trigonometrical ratios of 0°, 30°, 45°, 60° and 90°.

Indeed some calculators, when set to 'exact mode', will not give the approximate decimal value for sin 60°, cos 45°, tan 30° etc., but will instead give the exact values.

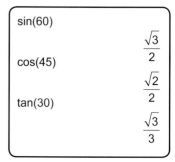

sin(60)
$$\frac{\sqrt{3}}{2}$$

cos(45)
$$\frac{\sqrt{2}}{2}$$

tan(30)
$$\frac{\sqrt{3}}{3}$$

*Question*: If it was not for this calculator display how would we know the exact value for sin 60°?

*Answer*: These exact values can be obtained by considering a number of specific triangles, as shown below.

The exact values for 30° and 60° can be obtained by considering an equilateral triangle ABC of side 2 units, as shown on the right.

Note that the perpendicular from A to BC will bisect BC and, by Pythagoras, will be of length $\sqrt{3}$ units.

It then follows that:

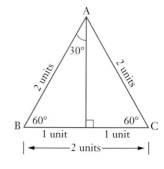

$$\sin 30° = \frac{1}{2} \qquad \cos 30° = \frac{\sqrt{3}}{2} \qquad \tan 30° = \frac{1}{\sqrt{3}}$$

$$\sin 60° = \frac{\sqrt{3}}{2} \qquad \cos 60° = \frac{1}{2} \qquad \tan 60° = \frac{\sqrt{3}}{1}$$

Notice that whilst the above statement gives the exact value of tan 30° as $\frac{1}{\sqrt{3}}$ the calculator shows this same value as $\frac{\sqrt{3}}{3}$, adopting the conventional style of displaying such values with denominators free of square roots (i.e. with a **rational** denominator).

ISBN 9780170390330

The exact values for 45° can be obtained from the diagram on the right showing an isosceles, right triangle DEF.

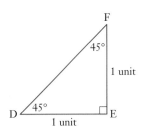

Note that DE and EF are each of unit length and so, by Pythagoras, DF will be of length $\sqrt{2}$ units.

Hence:    $\sin 45° = \dfrac{1}{\sqrt{2}}$     $\cos 45° = \dfrac{1}{\sqrt{2}}$     $\tan 45° = \dfrac{1}{1}$

The exact values for 0° and 90° can be obtained by considering the triangle GHI on the right.

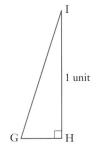

If point G is brought closer and closer to H, angle G will approach 90° and angle I will approach 0°. If IH is of unit length then, as angle G approaches 90°, IG will approach unit length and GH will approach zero length.

$$\sin 0° = \frac{0}{1} \qquad \cos 0° = \frac{1}{1} \qquad \tan 0° = \frac{0}{1}$$

$$\sin 90° = \frac{1}{1} \qquad \cos 90° = \frac{0}{1} \qquad \tan 90° = \frac{1}{0}$$

These exact values are summarised in the following table:

|  | 0° | 30° | 45° | 60° | 90° |
|---|---|---|---|---|---|
| **Sine** | 0 | $\dfrac{1}{2}$ | $\dfrac{1}{\sqrt{2}}$ | $\dfrac{\sqrt{3}}{2}$ | 1 |
| **Cosine** | 1 | $\dfrac{\sqrt{3}}{2}$ | $\dfrac{1}{\sqrt{2}}$ | $\dfrac{1}{2}$ | 0 |
| **Tangent** | 0 | $\dfrac{1}{\sqrt{3}}$ | 1 | $\sqrt{3}$ | undefined |

The reader is encouraged to learn these exact values and should be able to write them down without reference to the table.

The reader should also be able to determine exact values for the sine, cosine and tangent of a number of obtuse angles by making use of the following facts:

$$\sin \theta = \sin (180° - \theta) \qquad \cos \theta = -\cos (180° - \theta) \qquad \tan \theta = -\tan (180° - \theta)$$

(the last of these three statements following from the fact that $\tan \theta = \dfrac{\sin \theta}{\cos \theta}$).

## Exercise 1D

Without looking back at the table on the previous page, try to write down the exact values of the following and then check each one using a calculator.

**1** $\sin 0°$      **2** $\sin 30°$      **3** $\tan 45°$      **4** $\sin 60°$

**5** $\cos 60°$      **6** $\cos 0°$      **7** $\tan 90°$      **8** $\cos 45°$

**9** $\cos 30°$      **10** $\tan 60°$      **11** $\sin 90°$      **12** $\tan 0°$

**13** $\cos 60°$      **14** $\sin 45°$      **15** $\tan 30°$      **16** $\sin 120°$

**17** $\cos 135°$      **18** $\cos 150°$      **19** $\cos 120°$      **20** $\cos 180°$

**21** $\tan 135°$      **22** $\tan 120°$      **23** $\tan 150°$      **24** $\tan 180°$

**25** $\sin 180°$      **26** $\sin 150°$      **27** $\sin 135°$

Find the **exact** value of $x$ in each of the following.

**28**
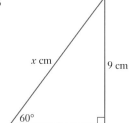

**29**
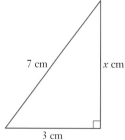

**30**
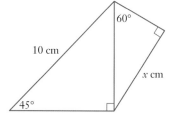

**31**
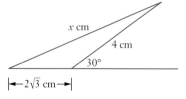

**32**
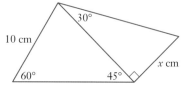

**33** For the diagram shown prove that

$$y = \frac{4\sqrt{2}\,(x)\sin\theta}{3}$$

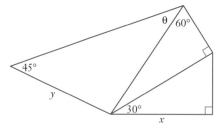

**34** For the diagram shown prove that

$$y = \frac{4\sin\theta}{\sin\phi}$$

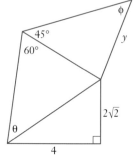

# The angle of inclination of a line

The **angle of inclination of a line** is the angle from 0° to 180° that the line makes with the *x*-axis, measured *from the x-axis, anticlockwise*.

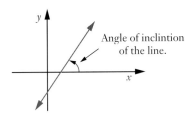

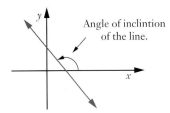

## Exercise 1E

For each of the lines shown below

    **a**    write down the angle of inclination of the line

    **b**    determine the gradient of the line as an exact value. (Think about it.)

**1**
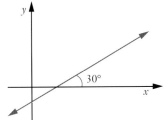

**2**
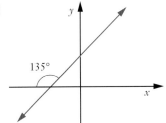

**3**
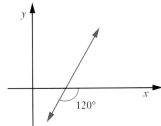

**4**
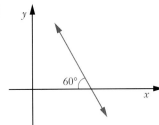

**5**
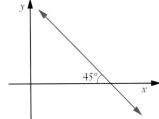

**6**
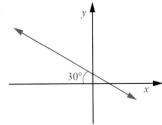

**7**  What is the relationship between the angle of inclination of a line and the gradient of that line?

# Miscellaneous exercise one

This miscellaneous exercise may include questions involving the work of this chapter and the ideas mentioned in the Preliminary work section at the beginning of the book.

**1** Expand and simplify:

**a** $2(3x + 4) + 5(x - 3)$      **b** $2(3x + 4) - 5(x - 3)$

**c** $5 + 2(5x - 4)$      **d** $5 - 2(5x - 4)$

**e** $2(3 - 4x) + 5(3x + 1)$      **f** $2(3 - 4x) - 5(3x + 1)$

**g** $(x + 3)(x + 5)$      **h** $(x + 3)(x - 5)$

**i** $(2x + 5)(x + 3)$      **j** $(2x - 5)(x - 3)$

**2** Factorise each of the following:

**a** $2x + 8$      **b** $6y + 9$

**c** $16ab + 12ac + 8a^2$      **d** $a^2 + a$

**e** $x^2 + 7x - 8$      **f** $x^2 - 9x + 8$

**g** $x^2 + 5x - 14$      **h** $x^2 - 8x + 12$

**i** $x^2 - 16$      **j** $2a^2 - 18$

**3** Simplify each of the following.

**a** $\sqrt{20}$      **b** $\sqrt{45}$

**c** $\sqrt{200}$      **d** $3\sqrt{5} \times 2\sqrt{5}$

**e** $\sqrt{15} \times \sqrt{3}$      **f** $6\sqrt{3} \times \sqrt{6}$

**g** $3\sqrt{5} \times 7\sqrt{2}$      **h** $\left(3\sqrt{2} + 1\right)^2$

**4** A ladder stands with its base on horizontal ground and its top against a vertical wall. When the base of the ladder is $a$ metres from the wall the ladder makes an angle of 80° with the ground. When the base of the ladder is pulled a further 20 cm from the wall the angle made with the ground becomes 75°. Find $a$ (correct to 2 decimal places) and the length of the ladder (correct to the nearest centimetre).

**5** From a lighthouse, ship A is 6.2 kilometres away on a bearing 040° and ship B is 10.8 km away on a bearing 100°. Find the distance and bearing of A from B.

**6** When the radius, $r$, of a circle increases then both the circumference, $C$, of the circle and the area, $A$, of the circle also increase. Does this mean that both $C$ and $r$, and $A$ and $r$, are in direct proportion?

**7** Ignoring any wastage needed for cutting, joining etc., what total length of steel would be needed to make twelve of the steel frameworks shown sketched on the right, rounding your answer up to the next ten metres.

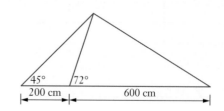

# 2.

# Radian measure

## Situation

A new machine is being designed and the design company is building a prototype. Part of the machine involves a continuous belt passing over two wheels, as shown in the diagram below (not drawn to scale).

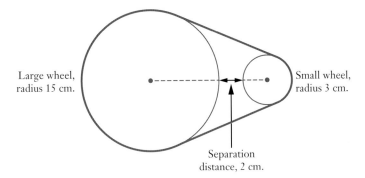

Large wheel, radius 15 cm.

Small wheel, radius 3 cm.

Separation distance, 2 cm.

The belt is to be made from a 'negligible stretch, high friction compound' and is made circular and then fitted exactly over the wheels. The belt is made by a computer controlled machine that only requires the operator to input the length of the radius of the circle and a circular belt of that radius will be produced.

Your task is to determine what the radius of the circular belt should be, giving your answer in centimetres correct to two decimal places.

## Arcs, sectors and segments

The situation above required you, amongst other things, to find the length of circular arcs. To determine the arc length you needed to determine the angle the arc subtended at the centre of the circle. Arc lengths, chord lengths, segment and sector areas can all be determined if we know the angle subtended at the centre of the circle and the radius of the circle.

If you have forgotten what terms like arc, segment and sector mean then the following diagrams should refresh your memory.

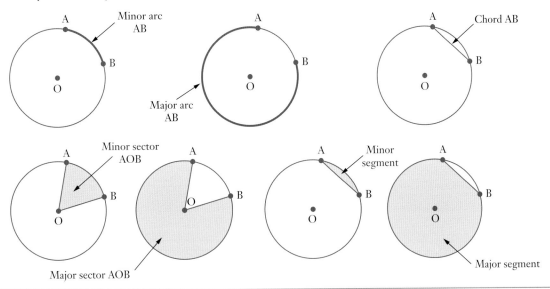

## EXAMPLE 1

Points A and B lie on the circumference of a circle centre O, radius 50 mm and are such that $\angle AOB = 40°$. Find the length of:

**a** minor arc AB, **b** chord AB.

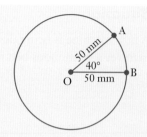

### Solution

**a** Circumference of circle $= 2 \times \pi \times 50$ mm

$$= 100\pi \text{ mm}$$

Length of minor arc AB $= \dfrac{40}{360} \times 100\pi$ mm

$$\approx 34.9 \text{ mm}$$

Minor arc AB is of length 35 mm (to nearest mm.)

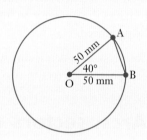

$$\frac{40}{360} \cdot 100\pi$$

$$\frac{100\pi}{9}$$

$$\frac{100\pi}{9} \blacktriangleright \text{Decimal}$$

$$34.90658504$$

Alternatively we could give the answer in exact form as $\dfrac{100\pi}{9}$ mm.

**b** By the cosine rule $AB^2 = 50^2 + 50^2 - 2(50)(50)\cos 40°$

$$\therefore AB \approx 34.2 \text{ cm}$$

The chord AB is of length 34 mm (to nearest mm.).

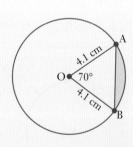

## EXAMPLE 2

Find the area of the segment shown shaded in the diagram.

### Solution

Area of circle $= \pi \times 4.1^2$ cm$^2$

$\therefore$ Area of minor sector AOB $= \dfrac{70}{360} \times \pi \times 4.1^2$ cm$^2$

Shaded area $= $ Area of sector AOB $-$ Area of $\triangle AOB$

$$= \frac{70}{360} \times \pi \times 4.1^2 - \frac{1}{2} \times 4.1 \times 4.1 \sin 70°$$

$$\approx 2.37 \text{ cm}^2$$

The shaded segment has an area of 2.4 cm$^2$ (correct to one decimal place).

ISBN 9780170390330

Find the area of the segment shown shaded in the diagram, giving your answer in exact form.

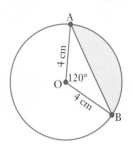

**Solution**

Area of circle $= \pi \times 4^2$

$\qquad = 16\pi \text{ cm}^2$

$\therefore$ Area of minor sector AOB $= \dfrac{120}{360} \times 16\pi \text{ cm}^2$

Shaded area $=$ Area of sector AOB $-$ Area of $\triangle$AOB

$\qquad = \dfrac{120}{360} \times 16\pi - \dfrac{1}{2} \times 4 \times 4 \sin 120°$

$\qquad = \left( \dfrac{16\pi}{3} - 4\sqrt{3} \right) \text{ cm}^2$

$$\dfrac{120}{360} \cdot 16 \cdot \pi - \dfrac{1}{2} \cdot 4 \cdot 4 \sin(120)$$

$$\dfrac{16 \cdot \pi}{3} - 4 \cdot \sqrt{3}$$

Shutterstock.com/Cmspic

## Exercise 2A

For numbers 1, 2 and 3 calculate the length of the arc AB shown in heavy type, giving each answer in centimetres and correct to one decimal place.

**1**
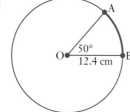

**2**
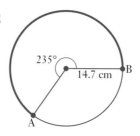

**3**
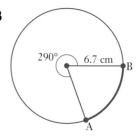

For numbers 4, 5 and 6 calculate the length of the arc AB shown in heavy type, giving each answer as an exact value.

**4**
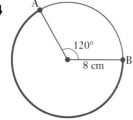

**5**
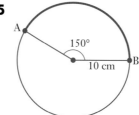

**6**
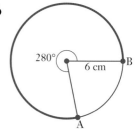

For numbers 7, 8 and 9 calculate the area of the shaded sector, giving each answer as an exact value.

**7**
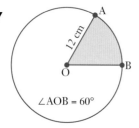

**8**
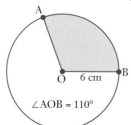

**9**
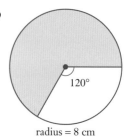

For numbers 10, 11 and 12 calculate the area of the shaded sector, giving each answer to the nearest square centimetre.

**10**
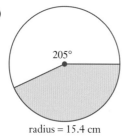

**11**
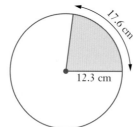

**12**
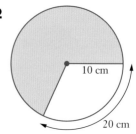

ISBN 9780170390330

For numbers 13, 14 and 15 calculate the area of the shaded segment, giving each answer to the nearest square centimetre.

**13**
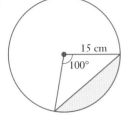
15 cm
100°

**14**
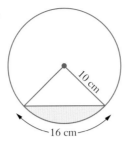
10 cm
16 cm

**15**
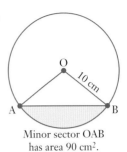
O
10 cm
A        B
Minor sector OAB
has area 90 cm².

For numbers 16, 17 and 18 calculate the area of the shaded segment, giving each answer as an exact value.

**16**
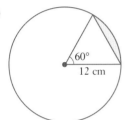
60°
12 cm

**17**
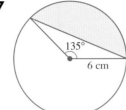
135°
6 cm

**18**
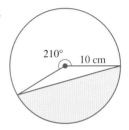
210°   10 cm

**19** A and B are two points on the circumference of a circle centre O and radius 15.2 cm.
If ∠AOB = 112° find the length of

   **a**   the minor arc AB

   **b**   the major arc AB.

**20** A and B are two points on the circumference of a circle centre O. Angle AOB is of size 75° and the minor arc AB is of length 24 cm. Calculate the radius of the circle in centimetres, correct to one decimal place.

**21** Points A and B lie on the circumference of a circle centre O, radius 15 cm. Find the area of the minor sector AOB given that ∠AOB = 50°.

**22** A and B are two points on the circumference of a circle centre O and radius 18 cm. Find the area of the minor segment cut off by the chord AB given that ∠OAB = 20°.

**23** Find the size of the acute angle AOB, correct to the nearest degree, given that A and B are two points on the circumference of a circle centre O, radius 12 cm, and the major sector AOB has an area of 378 cm².

**24** Find the area of the minor segment cut off by a chord of length 10 cm drawn in a circle of radius 12 cm, giving your answer in square centimetres correct to one decimal place.

**25** A clock has an hour hand of length 8 cm and a minute hand of length 12 cm.
Calculate the distance travelled by the tip of each hand in half an hour.
(Give your answers in exact form.)

**26** One nautical mile is defined to be the distance on the surface of the Earth that subtends an angle of one minute at the centre of the Earth (1 degree = 60 minutes). How many nautical miles are travelled by a ship travelling due north and changing its latitude from 5° N to 8° N?

Assuming the Earth to be a sphere of radius 6 350 km express one nautical mile in kilometres, correct to two decimal places.

**27** A minor sector is removed from a circular piece of card (see Figure 1).

Figure 1

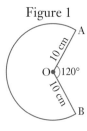

Figure 2

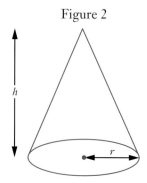

By joining OA to OB the remaining major sector forms a conical hat (see Figure 2). Find $h$ and $r$, the height and base radius of the hat respectively.

# Radians

Consider the concentric circles shown on the right. Lines OA and OB are drawn from the common centre O (see diagram). The minor arcs $A_1B_1$, $A_2B_2$, $A_3B_3$ and $A_4B_4$ each subtend an angle $\theta°$ at O and will be of increasing length.

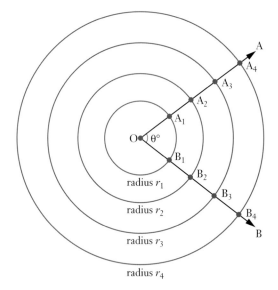

$$A_1B_1 = \frac{\theta}{360} \times 2\pi r_1$$

$$A_2B_2 = \frac{\theta}{360} \times 2\pi r_2$$

$$A_3B_3 = \frac{\theta}{360} \times 2\pi r_3$$

$$A_4B_4 = \frac{\theta}{360} \times 2\pi r_4$$

Note that the ratio of arc length, $A_nB_n$, to radius, $r_n$, is constant:

$$\frac{A_1B_1}{r_1} = \frac{A_2B_2}{r_2} = \frac{A_3B_3}{r_3} = \frac{A_4B_4}{r_4} = \frac{\theta}{360} \times 2\pi$$

This ratio of arc length to radius can be used as a measure of angle and gives us an alternative unit for measuring angles.

This unit is called a **radian** and proves to be a very useful measure of angle for advanced mathematics. If the ratio of arc length to radius is equal to 1 the angle subtended at the centre is 1 radian.

ISBN 9780170390330

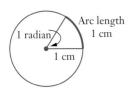
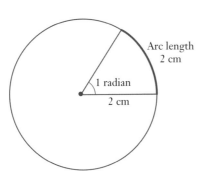
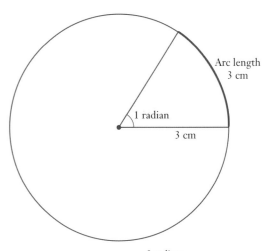

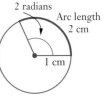

An arc of length 1 unit, in a circle of unit radius, subtends an angle of 1 radian at the centre of the circle.

An arc of length 2 units, in a circle of unit radius, subtends an angle of 2 radians at the centre of the circle, and so on.

# Radians ↔ degrees

An arc of length 1 unit, in a circle of unit radius, subtends an angle of 1 radian at the centre of the circle. Thus an arc of length $2\pi$ units, in a circle of unit radius, will subtend an angle of $2\pi$ radians at the centre of the circle. However, if the radius is 1 unit an arc of $2\pi(1)$ is the full circumference of the circle and will subtend an angle of 360° at the centre.

Converting degrees and radians

Thus $\qquad\qquad\qquad 2\pi \text{ radians} = 360°$

i.e. $\qquad\qquad\qquad \pi \text{ radians} = 180°$

Thus, correct to one decimal place, 1 radian is equivalent to 57.3°.

---

**EXAMPLE 4**

Convert to radians, leaving your answers in terms of $\pi$.

**a**   60° $\qquad\qquad$ **b**   90° $\qquad\qquad$ **c**   125°

**Solution**

**a** $\qquad 180° = \pi \text{ radians}$

$\therefore \quad 1° = \dfrac{\pi}{180} \text{ radians}$

$\therefore \quad 60° = \dfrac{\pi}{180} \times 60$

$\qquad\quad = \dfrac{\pi}{3} \text{ radians}$

**b** $\qquad 180° = \pi \text{ radians}$

$\therefore \quad 1° = \dfrac{\pi}{180} \text{ radians}$

$\therefore \quad 90° = \dfrac{\pi}{180} \times 90$

$\qquad\quad = \dfrac{\pi}{2} \text{ radians}$

**c** $\qquad 180° = \pi \text{ radians}$

$\therefore \quad 1° = \dfrac{\pi}{180} \text{ radians}$

$\therefore \quad 125° = \dfrac{\pi}{180} \times 125$

$\qquad\quad = \dfrac{25\pi}{36} \text{ radians}$

---

Note:

- To convert degrees ↔ radians we use the exact result π radians = 180° rather than the approximation 1 radian ≈ 57.3°.

- When an angle is given in radians the word radian is optional. The answer to **Example 4a** could be given as $\frac{\pi}{3}$ radians or simply as $\frac{\pi}{3}$.

  When an angle is given with no units stated then the angle should be assumed to be in radians.

- Knowing conversions such as $\frac{\pi}{3}$ radians = 60°, it follows that $\sin \frac{\pi}{3} = \frac{\sqrt{3}}{2}$.

## EXAMPLE 5

Express the following in degrees, correct to the nearest degree if rounding is necessary.

a   $\frac{\pi}{8}$ radians

b   2.3 radians

**Solution**

a   π radians = 180°

$$\therefore \frac{\pi}{8} \text{ radians} = \frac{180}{8} \text{ degrees}$$

$$= 22.5°$$

b   π radians = 180°

$$\therefore \quad 1 \text{ radian} = \frac{180}{\pi} \text{ degrees}$$

$$\therefore \quad 2.3 \text{ radians} = \frac{180}{\pi} \times 2.3$$

$$= 132° \text{ to nearest degree}$$

- Explore the capability of your calculator to change between the various units for measuring angle.

The next example shows that the trigonometrical ratios can still be applied with angles given in radians. We do not need to change the angles to degrees but instead set our calculator to read angles as radians.

## EXAMPLE 6

Find the length of side AB as shown in the diagram on the right.

**Solution**

Let AB be of length $x$ cm.

$$\tan 1.2 = \frac{x}{5.2}$$

$$\therefore \quad x = 5.2 \tan 1.2$$

$$\approx 13.38$$

The side AB is of length 13.4 cm, correct to one decimal place.

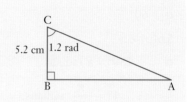

## Exercise 2B

For numbers 1 to 6 state the size of angle θ in radians.

**1**

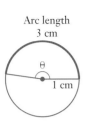

Arc length
3 cm

θ

1 cm

**2**

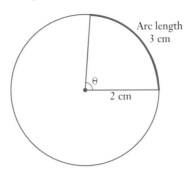

Arc length
3 cm

θ

2 cm

**3**

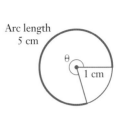

Arc length
5 cm

θ

1 cm

**4**

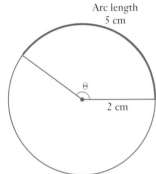

Arc length
5 cm

θ

2 cm

**5**

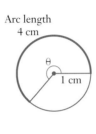

Arc length
4 cm

θ

1 cm

**6**

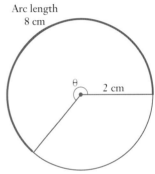

Arc length
8 cm

θ

2 cm

Express the following in radians as exact values in terms of π.

**7** 90°

**8** 30°

**9** 150°

**10** 135°

**11** 5°

**12** 18°

**13** 80°

**14** 130°

Express the following as degrees.

**15** $\frac{\pi}{4}$ rads

**16** $\frac{\pi}{3}$ rads

**17** $\frac{2\pi}{3}$ rads

**18** π rads

**19** $\frac{\pi}{12}$ rads

**20** $\frac{\pi}{5}$ rads

**21** $\frac{7\pi}{36}$ rads

**22** $\frac{7\pi}{18}$ rads

Express the following as radians, correct to two decimal places.

**23** 32°

**24** 63°

**25** 115°

**26** 170°

**27** 16°

**28** 84°

**29** 104°

**30** 26°

Change the following to degrees giving answers correct to the nearest degree.

**31** 1.5 rads　　　　**32** 2.3 rads　　　　**33** 1.4 rads　　　　**34** 0.6 rads

Without using a calculator state the exact values of the following.

**35** $\sin \dfrac{\pi}{4}$　　　　**36** $\sin \dfrac{5\pi}{6}$　　　　**37** $\cos \dfrac{3\pi}{4}$　　　　**38** $\sin \dfrac{\pi}{2}$

**39** $\sin \dfrac{2\pi}{3}$　　　　**40** $\sin \dfrac{3\pi}{4}$　　　　**41** $\cos \dfrac{\pi}{4}$　　　　**42** $\tan \dfrac{2\pi}{3}$

**43** $\cos \dfrac{\pi}{2}$　　　　**44** $\tan \dfrac{\pi}{2}$　　　　**45** $\cos \dfrac{2\pi}{3}$　　　　**46** $\tan \dfrac{5\pi}{6}$

**47** $\cos \dfrac{5\pi}{6}$　　　　**48** $\tan \pi$　　　　**49** $\cos \dfrac{\pi}{3}$　　　　**50** $\sin \pi$

Use your calculator to determine the following correct to two decimal places.

**51** $\sin 1$　　　　**52** $\cos 2$　　　　**53** $\tan 2.5$　　　　**54** $\sin 3$

**55** $\cos 0.6$　　　　**56** $\cos 0.15$　　　　**57** $\tan 1.3$　　　　**58** $\sin 2.3$

Find the *acute* angle $\theta$ in each of the following giving your answers in radians correct to two decimal places.

**59** $\sin \theta = 0.2$　　　　**60** $\cos \theta = 0.2$　　　　**61** $\tan \theta = 0.35$　　　　**62** $\tan \theta = 1.7$

**63** Convert the following angular speeds to radians/second.

　　**a** 3 revolutions/second　　　**b** 15 revolutions/minute　　　**c** 90 degrees/second

**64** Convert the following angular speeds to revolutions/minute.

　　**a** $2\pi$ radians/minute　　　**b** $\dfrac{3\pi}{4}$ radians/second　　　**c** $\dfrac{\pi}{3}$ radians/second.

Find the value of $x$ in each of the following, giving your answers correct to one decimal place.

**65**

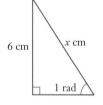

**66**

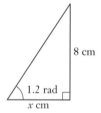

**67**

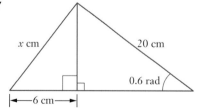

**68**

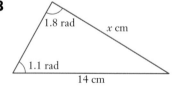

**69**

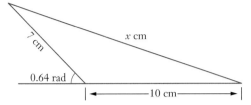

7 cm
x cm
0.64 rad
10 cm

**70**

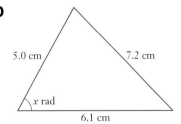

5.0 cm
7.2 cm
x rad
6.1 cm

**71** Through what angle, in radians, does the minute hand of a clock rotate in

**a** 15 minutes? **b** 40 minutes? **c** 50 minutes? **d** 55 minutes?

**72** A grad is another unit that can be used to measure angles. One right angle = 100 grads. Convert the following to radians.

**a** 50 grads **b** 75 grads **c** 10 grads **d** 130 grads

**73** A simple gauge is to be made for measuring the diameter of pipes. The gauge will be in the form of a rectangular piece of wood from which a V-shape has been cut. The V is then placed on the pipe and the point of contact, D, (see diagram) allows the diameter to be read directly from the graduations on AB.

The V shape is cut such that ∠BAC = 1 radian and AB = AC = 12 cm.

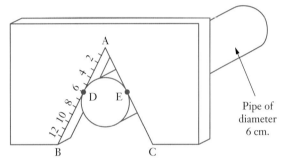

A
D   E
12 10 8 6 4 2
B       C

Pipe of diameter 6 cm.

**a** Draw a line 12 cm long to represent BA. Calibrate it so that diameters from 1 cm to 12 cm could be read **directly** from the point of contact.

**b** Would calibration have been simpler if $\angle BAC = \dfrac{\pi}{2}$ radians? Explain your answer.

# Arcs, sectors and segments revisited

Let us now consider again arc lengths, sector areas and segment areas but this time let the angle subtended at the centre of the circle be in radians.

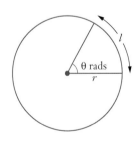

θ rads
r
l

The central angle, in radians, is given by $\dfrac{\text{arc length}}{\text{radius}}$.

Thus $\theta = \dfrac{l}{r}$ with $\theta$ and $r$ as defined in the diagram on the right.

Thus $l = r\theta$.

$$\text{Arc length} = r\theta$$

Remembering that 1 revolution is $2\pi$ radians it follows that, with $\theta$ as shown in the diagram on the right,

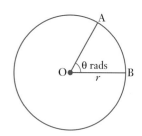

$$\frac{\theta}{2\pi} = \frac{\text{Area of sector AOB}}{\pi r^2}$$

$$\therefore \quad \text{Sector area} = \frac{\theta}{2\pi} \times \pi r^2$$

$$= \frac{1}{2} r^2 \theta$$

Thus

$$\boxed{\text{Sector area} = \frac{1}{2} r^2 \, \theta}$$

From the diagram on the right we see that

Shaded area = Area of sector AOB – Area of $\triangle$AOB

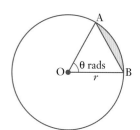

$$= \frac{1}{2} r^2 \theta - \frac{1}{2} r^2 \sin \theta$$

$$= \frac{1}{2} r^2 (\theta - \sin \theta)$$

Thus

$$\boxed{\text{Segment area} = \frac{1}{2} r^2 (\theta - \sin \theta)}$$

## EXAMPLE 7

Calculate the area of the shaded region in each of the following diagrams.

a
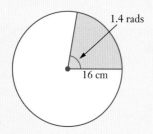

b
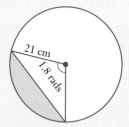

**Solution**

a    Sector area $= \dfrac{1}{2} r^2 \theta$

$$= \frac{1}{2} \times 16^2 \times 1.4$$

$$= 179.2 \text{ cm}^2$$

The shaded region has area 179 cm². (To the nearest cm².)

b    Area $= \dfrac{1}{2} r^2 (\theta - \sin \theta)$

$$= \frac{1}{2} \times 21^2 \times (1.8 - \sin 1.8)$$

$$\approx 182.2 \text{ cm}^2$$

The shaded region has area 182 cm². (To the nearest cm².)

ISBN 9780170390330

## EXAMPLE 8

Points A and B are points on the circumference of a circle, centre O and radius 4 cm. If the minor arc AB is of length 10 cm find the area of the minor sector AOB.

### Solution

First draw a diagram:

Arc length,     $l = r\theta$

Thus            $10 = 4\theta$

and so          $\theta = 2.5$

$\therefore$    Sector area $= \dfrac{1}{2}(4)^2(2.5)$

$= 20$ cm$^2$

The minor sector AOB has an area of 20 cm$^2$.

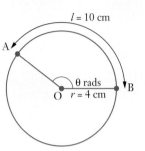

## Exercise 2C

Find the lengths of the arcs shown by heavy type in the following diagrams.

**1**

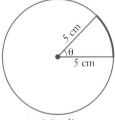

$\theta = 0.8$ radians

**2**

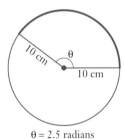

$\theta = 2.5$ radians

**3**

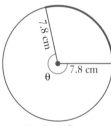

$\theta = 4.5$ radians

Find the areas of the sectors shown shaded in the following diagrams.

**4**

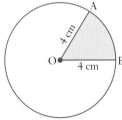

$\angle AOB = 1$ radian

**5**

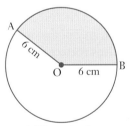

$\angle AOB = 2.5$ radians

iStock.com/oytun karadayi

**6**

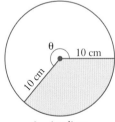

$\theta = 4$ radians

Find the areas of the segments shown shaded in the following diagrams.

**7**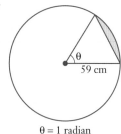

θ
59 cm

θ = 1 radian

**8**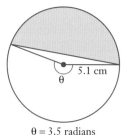

5.1 cm
θ

θ = 3.5 radians

**9**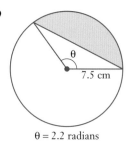

θ
7.5 cm

θ = 2.2 radians

**10** Find the length of the arc AB given that it subtends an angle of 1.2 radians at the centre of the circle of which it forms a part and the radius of the circle is 15 cm.

**11** Points A and B lie on the circumference of a circle, centre O, radius 15 cm. If the minor arc AB subtends an angle of 0.8 radians at O find the area of

**a** the minor sector OAB,

**b** the major sector OAB (to the nearest square centimetre).

**12** A and B are two points on a circle centre O and radius 8 cm. If arc AB subtends an angle of 1 radian at O find

**a** the length of the minor arc AB,

**b** the area of that part of the minor sector OAB not lying in triangle OAB. (Give your answer in square centimetres correct to one decimal place.)

**13** A and B lie on the circumference of a circle centre O and radius 5 cm. The minor sector OAB has an area of 15 cm².

**a** Calculate the length of the minor arc AB.

**b** Calculate the area of the minor segment cut off by the chord AB. (Give your answer in square centimetres correct to two decimal places).

**14** Points A and B lie on the circumference of a circle, centre at point O and with radius 8 cm. If the minor arc AB is of length 20 cm find the area of the minor sector OAB.

**15** Points A and B lie on the circumference of a circle, centre O and of radius 6 cm. If the minor sector OAB has an area of 9 cm² find the area of the minor segment cut off by the chord AB. (Giving your answer in square centimetres correct to two decimal places.)

**16** Find the area of the shaded region shown given that O is the centre of both circles, OD = DC = 6 cm and θ = 1.5 radians.

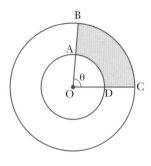

B
A
θ
O D C

ISBN 9780170390330

**17** Find the area of the shaded region shown given that O is the centre of both circles, OA = 5 cm, AB = 4 cm and $\theta$ = 3 radians.

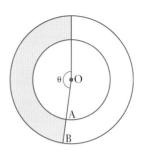

**18** The diagram shows two overlapping circles with the region common to both circles shown shaded.

Find the area of this shaded region given that $\angle O_1AO_2$ and $\angle O_1BO_2$ are right angles. (Give your answer in square centimetres, and correct to one decimal place.)

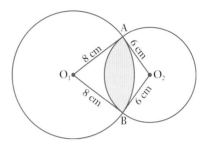

**19** The diagram shows two circles with common centre O. The region shown shaded is bounded by the minor arc BC, the chord AD and the lines CD and BA. Calculate the area of this region given that OA = 5 cm, AB = 3 cm and $\angle AOD$ = 0.8 radians. (Give your answer in square centimetres, and correct to one decimal place.)

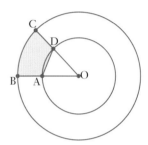

**20** The diagram shows the two tangents drawn from the point C to a circle centre O and radius 6 cm, touching the circle at the points A and B.

Find the area of the region shown shaded. (The angle between a tangent and the radius drawn at the point of contact is a right angle.) (Give your answer in square centimetres and correct to one decimal place.)

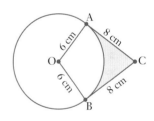

**21** Calculate the area of the region shown shaded given that $\angle AOB$ = 0.5 radians, $\angle BOC$ = 1 radian, $\angle COD$ = 0.5 radians and the circle is of radius 5 cm. (Give your answer in square centimetres and correct to two decimal places.)

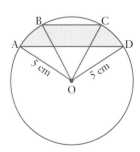

**22** A pendulum AB has end *A* fixed and a weight attached at B. In one swing the weight travels from B to C and back again (see diagram).

The pendulum is of length 75 cm and ∠BAC = 0.8 radians.

**a** How far does the weight travel in one swing?

**b** By how much does the length of the arc BC exceed that of the chord BC? (Answer to the nearest millimetre.)

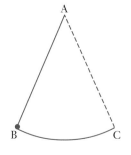

**23** Find, to the nearest 10 mm², the area of the shaded region shown given that $O_1O_2 = 70$ mm.

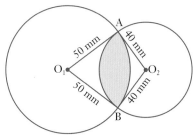

**24** Two circles of radius 10 cm and 7 cm have their centres 15 cm apart.

Find the perimeter of the region common to both circles, giving your answer in centimetres and correct to one decimal place.

**25** What percentage of the circumference of a circular disc of radius 10 cm can be illuminated from a point source of light in the plane of the disc and 12 cm from it (see diagram)? Give your answer to the nearest percent.

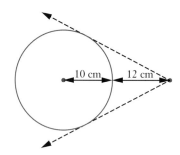

**26** A goat is tethered to a post by a rope that is ten metres long. The goat is able to graze over any area that the rope allows it to reach other than that excluded by a straight fence. The perpendicular distance from the post to the fence is 6 m. Over what area can the goat graze (to the nearest square metre)?

**27** A goat is tethered to a post by a rope that is twelve metres long. If the post is situated half way between two parallel fences that are ten metres apart, over what area can the goat graze (to the nearest square metre)?

ISBN 9780170390330

**28** Find, to the nearest centimetre, the length of the continuous belt passing around two wheels as shown in the diagram on the right.

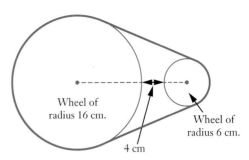

Wheel of radius 16 cm.

4 cm

Wheel of radius 6 cm.

**29** A door is to be made to the specifications shown in the diagram i.e. a circular segment on top of a rectangle.

The top segment is part of a circle having its centre at the intersection of the diagonals of the rectangle ABCD. If AB = 80 cm and AD = 200 cm find the area of the door correct to the nearest ten square centimetres.

**30** Find, to the nearest centimetre, the length of the continuous belt passing around the three wheels as shown in the diagram (not drawn to scale).

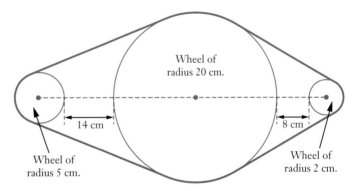

Wheel of radius 20 cm.

14 cm

8 cm

Wheel of radius 5 cm.

Wheel of radius 2 cm.

**31 a** A minor sector of a circle has perimeter 14 cm and area 10 cm$^2$. Find the radius of the circle.

 **b** A major sector of a circle has perimeter 14 cm and area 10 cm$^2$. Find the radius of the circle.

**32** Circles of radius 10 cm and 5 cm touch each other tangentially and both touch the line AB (see diagram).

Find the area of the region shown shaded in the diagram. (Answer to nearest 0.1 cm$^2$.)

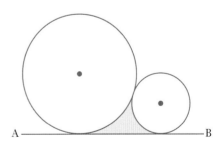

A

B

**33** Triangle ABC has AB = 7 cm, AC = 6 cm and BC = 5 cm. Three circles are drawn, one with centre A and radius 4 cm, one with centre B and radius 3 cm and one with centre C and radius 2 cm. What percentage of the area of the triangle fails to lie in any of the circles? (Answer to nearest 0.1%.)

# Miscellaneous exercise two

This miscellaneous exercise may include questions involving the work of this chapter, the work of any previous chapters, and the ideas mentioned in the Preliminary work section at the beginning of the book.

**1** Without the assistance of a calculator, expand each of the following.

    **a**   $(x + 3)(2x - 1)$                  **b**   $(x + 7)(3x - 4)$

    **c**   $(x + 5)(x - 1)(x + 3)$        **d**   $(2x + 1)(x - 3)(x - 2)$

**2** Simplify each of the following by expressing them as equivalent statements without any square roots in the denominators (i.e. rationalise the denominators).

    **a**  $\dfrac{1}{\sqrt{2}}$        **b**  $\dfrac{1}{\sqrt{3}}$        **c**  $\dfrac{5}{\sqrt{2}}$        **d**  $\dfrac{6}{\sqrt{3}}$

    Hint for **e** to **h**: To simplify $\dfrac{a}{b+\sqrt{c}}$ multiply by 1, written in the form $\dfrac{b-\sqrt{c}}{b-\sqrt{c}}$.

    **e**  $\dfrac{1}{3+\sqrt{5}}$      **f**  $\dfrac{1}{3-\sqrt{2}}$      **g**  $\dfrac{2}{1+\sqrt{5}}$      **h**  $\dfrac{3}{\sqrt{5}+\sqrt{2}}$

**3** From a point A, level with the base of the town hall, the angle of elevation of the topmost point of the building is 35°. From point B, also at ground level but 30 metres closer to the hall, the same point has an angle of elevation of 60°. Find how high the topmost point is above ground level. (Give your answer correct to the nearest metre.)

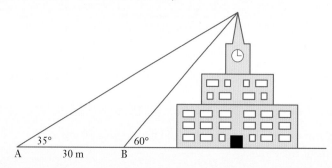

**4** A playground roundabout of radius 1.8 m makes one revolution every five seconds. Find, to the nearest centimetre, the distance travelled by a point on the roundabout in one second if the point is

    **a**   1.8 m from the centre of rotation       **b**   1 m from the centre of rotation.

**5** From a lighthouse, ship A is 17.2 km away on a bearing S 60° E and ship B is 14.1 km away on a bearing N 80° W. How far, and on what bearing, is B from A?

**6** The diagram on the right shows the sketch made by a surveyor after taking measurements for a block of land ABCD.

Find the area and the perimeter of the block.

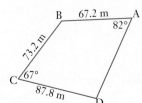

# 3.

# Function

- So exactly what do we mean by 'a function'?
- Machine analogy
- Natural domain of a function
- Miscellaneous exercise three

# So exactly what do we mean by 'a function'?

Five soccer players took part in a penalty taking competition. Each player took eight penalties and the arrow diagram on the right shows the number of goals each player scored from these eight penalties.

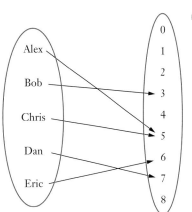

If we select a name from the first set, the arrow diagram indicates the number of goals that player scored.

As was mentioned in the *Preliminary work* section at the beginning of this text, any relationship that assigns to each element of one set one *and only one* element from a second set is called a **function**.

We say that an element of the first set **maps onto** an element of the second set. For example, Alex in the first set maps onto 5 in the second set. We write Alex $\mapsto$ 5.

The first set we call the **domain**, the second set we call the **co-domain** and those elements of the co-domain that the elements of the first set map onto form the **range**.

Thus in the above function,      the domain is    {Alex, Bob, Chris, Dan, Eric},
the co-domain is    {0, 1, 2, 3, 4, 5, 6, 7, 8}
and the range is    {3, 5, 6, 7}.

Notice that in this function two elements of the domain, Alex and Chris, map onto the same element of the range, 5. We call such functions, in which more than one element of the domain map onto the same element of the range, **many-to-one** functions.

If each element of the domain is mapped onto a different element of the range then the function is said to be **one-to-one**.

One-to-many relationships can occur but under our requirement that a function takes one element from the domain and assigns to it one *and only one* element from the range, a one-to-many relationship would not be called a function. (Thus the arrow diagram for a function cannot have any elements in the first set from which more than one arrow leaves.) This terminology is further illustrated below:

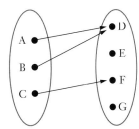

A many-to-one function
Domain {A, B, C}
Co-domain {D, E, F, G}
Range {D, F}

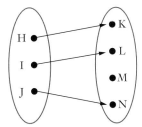

A one-to-one function
Domain {H, I, J}
Co-domain {K, L, M, N}
Range {K, L, N}

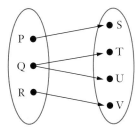

A 'one to many' relationship
Therefore not a function.

Many mathematical functions can be formed using a calculator. The correct key strokes can perform certain functions on a given number and give the appropriate output. For example:

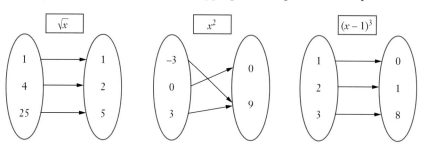

These functions may be *one-to-one*, as in $\sqrt{x}$ and $(x-1)^3$ shown above, or they may be *many-to-one*, as in $x^2$ shown above, for which $(-3)^2$ and $(3)^2$ have the same output, 9.

The domain and range of most of the functions we will deal with will be sets of numbers.

For example:

The '× 2' function
with domain {1, 2, 3}

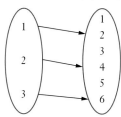

Range {2, 4, 6}.

We write:
$f(1) = 2$
$f(2) = 4$
$f(3) = 6$

Thus $f(x) = 2x$

The '× 2 and subtract 1' function
with domain {1, 2, 3}

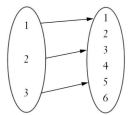

Range {1, 3, 5}.

We write:
$f(1) = 1$
$f(2) = 3$
$f(3) = 5$

Thus $f(x) = 2x - 1$

When several *functions* are used in one question, $g(x)$ and $h(x)$ are commonly used to distinguish between them.

## EXAMPLE 1

If $f(x) = 5x + 1$ and $g(x) = x^2 - 5$ determine  **a** $f(3)$  **b** $g(-2)$

**c** $p$ given that $f(p) = g(p)$

### Solution

**a**
$f(x) = 5x + 1$
$\therefore f(3) = 5(3) + 1$
$= 16$
Thus $f(3)$ is 16.

**b**
$g(x) = x^2 - 5$
$\therefore g(-2) = (-2)^2 - 5$
$= -1$
Thus $g(-2)$ is –1.

**c**
$f(p) = g(p)$
$\therefore 5p + 1 = p^2 - 5$
$0 = p^2 - 5p - 6$
$0 = (p + 1)(p - 6)$
$p = -1$ or 6

ISBN 9780170390330

# Machine analogy

As the *Preliminary work* reminded us it can be useful at times to consider a function as a machine. A box of numbers (the domain) is fed into the machine, a certain rule is applied to each number, and the resulting output forms a new box of numbers, the range.

In this way $f(x) = x^2 + 3$, with domain $\{1, 2, 3, 4, 5\}$, could be 'pictured' as follows:

# Natural domain of a function

If we are not given a specific domain we assume it to be all the numbers which the function 'can cope with', i.e. all of the real numbers for which the function is *defined*. We call this the **natural domain** or **implied domain** of the function.

For example $\qquad f(x) = 2x + 3 \qquad$ is defined for all real $x$.

Thus $\qquad\qquad f(x) = 2x + 3 \qquad$ has natural domain $\mathbb{R}$.

(As mentioned in the preliminary work section we use $\mathbb{R}$, or **R**, to represent the set of all real numbers.)

$$g(x) = \frac{1}{x-3} \qquad \text{is not defined for } x = 3.$$

Thus $\qquad\qquad g(x) = \dfrac{1}{x-3} \qquad$ has natural domain $\{x \in \mathbb{R}: x \neq 3\}$.

Reading '$\in$' as 'is a member of' and ':' as 'such that', $\{x \in \mathbb{R}: x \neq 3\}$ can be read as *x is a member of the set of real numbers such that x is not equal to 3*.

$$h(x) = \sqrt{x-3} \qquad \text{is only defined for } x - 3 \geq 0$$

Thus $\qquad\qquad h(x) = \sqrt{x-3} \qquad$ has natural domain $\{x \in \mathbb{R}: x \geq 3\}$.

---

### EXAMPLE 2

State the range of each of the following functions for the given domain.

**a** $\quad f(x) = x + 1 \; \{x \in \mathbb{R}: 2 \leq x \leq 5\}$ 
**b** $\quad f(x) = \sqrt{x}$ for the natural domain of $f(x)$.

#### Solution

**a** Adding 1 to all the real numbers from 2 to 5 will give all the real numbers from 3 to 6. Thus the range is $\{y \in \mathbb{R}: 3 \leq y \leq 6\}$.

Note that we could use any letter to define the range but in this book we will tend to use $x$ as the variable when defining a domain and $y$ as the variable when defining a range.

**b** The natural domain of the function is $\{x \in \mathbb{R}: x \geq 0\}$. 
This function could then output any non negative real number. 
Thus the range is $\{y \in \mathbb{R}: y \geq 0\}$.

---

Alternatively Example 2 could be considered graphically. If we view the graph $y = f(x)$ then the required range will be the $y$ values corresponding to the $x$ values in the domain.

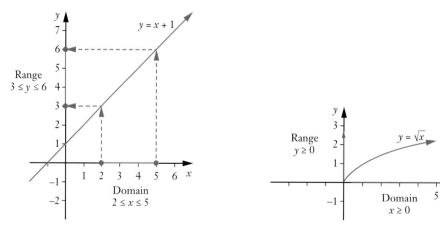

Note that with our requirement that a function takes one element from the domain and assigns to it one *and only one* element from the range, the graph of a function must be such that:

If a vertical line is moved from the left of the domain to the right it must never cut the graph in more than one place.

This is called the **vertical line test**.

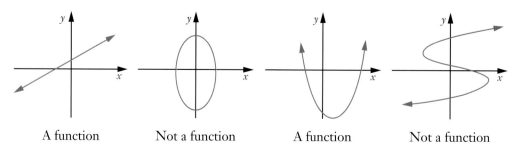

| A function | Not a function | A function | Not a function |

Note: We could use a similar *horizontal line test* to determine whether a function is a one-to-one function or not.

## Exercise 3A

**1** Which of the following arrow diagrams show functions?

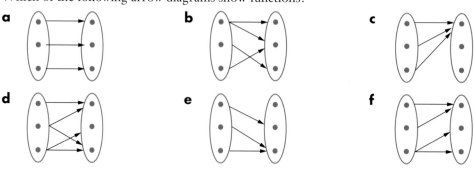

a      b      c

d      e      f

**2** Which of the following shows the graph of a function?

**a**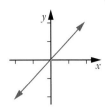

**b**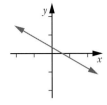

**c**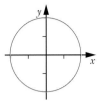

**d**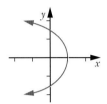

**e**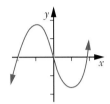

**f**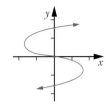

**3** State the range of each of the following function machines for the domains shown.

**a**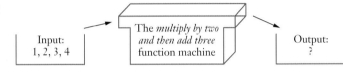

Input: 1, 2, 3, 4 → The *multiply by two and then add three* function machine → Output: ?

**b**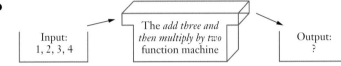

Input: 1, 2, 3, 4 → The *add three and then multiply by two* function machine → Output: ?

**c**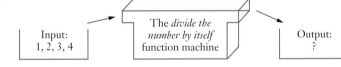

Input: 1, 2, 3, 4 → The *divide the number by itself* function machine → Output: ?

**d**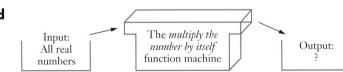

Input: All real numbers → The *multiply the number by itself* function machine → Output: ?

**4** For $f(x) = 5x - 2$ determine each of the following.

 **a** $f(4)$    **b** $f(-1)$    **c** $f(3)$    **d** $f(1.2)$

 **e** $f(3) + f(2)$    **f** $f(5)$    **g** $f(-5)$    **h** $f(a)$

 **i** $f(2a)$    **j** $f(a^2)$    **k** $3f(2)$    **l** $f(a + b)$

 **m** The value of $p$ for which $f(p) = 33$.

 **n** The value of $q$ for which $f(q) = -12$.

**5** For $f(x) = 4x - 7$, $g(x) = x^2 - 12$ and $h(x) = x^2 - 3x + 3$, determine each of the following.

    **a**   $f(4)$              **b**   $f(0)$              **c**   $g(3)$              **d**   $g(-3)$

    **e**   $h(-5)$           **f**   $h(5)$             **g**   $h(-2)$          **h**   $3f(a)$

    **i**   $f(3a)$            **j**   $3g(a)$           **k**   $g(3a)$

    **l**   The values of $p$ for which $g(p) = 24$.

    **m**   The value of $q$ for which $g(q) = h(q)$.

    **n**   The values of $r$ for which $h(r) = f(r) + 28$.

**6** Which numbers can each of the following functions **not** cope with? (i.e. which numbers must not be included in the domain?)

    **a**   $f(x) = \sqrt{x-1}$     **b**   $f(x) = x^2 + 1$     **c**   $f(x) = \dfrac{1}{x}$     **d**   $f(x) = \dfrac{1}{1-x}$

**7** Which numbers is it impossible for each of the following functions to output? (i.e. which numbers will not be included in the range?)

    **a**   $f(x) = \sqrt{x-1}$     **b**   $f(x) = x^2 + 1$     **c**   $f(x) = \dfrac{1}{x}$     **d**   $f(x) = \dfrac{1}{1-x}$

For questions **8** to **22** state the range of each function for the given domain.

**8**   Function:    $f(x) = x + 5$,             Domain:      $\{x \in \mathbb{R}: 0 \le x \le 3\}$

**9**   Function:    $f(x) = x - 3$,             Domain:      $\{x \in \mathbb{R}: 0 \le x \le 3\}$

**10**   Function:    $f(x) = 3x$,               Domain:      $\{x \in \mathbb{R}: -2 \le x \le 5\}$

**11**   Function:    $f(x) = 4x$,               Domain:      $\{x \in \mathbb{R}: 5 \le x \le 10\}$

**12**   Function:    $f(x) = 2x - 1$,           Domain:      $\{x \in \mathbb{R}: 0 \le x \le 5\}$

**13**   Function:    $f(x) = 1 - x$,            Domain:      $\{x \in \mathbb{R}: 0 \le x \le 5\}$

**14**   Function:    $f(x) = x^2$,              Domain:      $\{x \in \mathbb{R}: -1 \le x \le 3\}$

**15**   Function:    $f(x) = (x + 1)^2$,        Domain:      $\{x \in \mathbb{R}: -2 \le x \le 3\}$

**16**   Function:    $f(x) = x^2 + 1$,          Domain:      $\{x \in \mathbb{R}: -1 \le x \le 3\}$

**17**   Function:    $f(x) = \dfrac{1}{x}$,            Domain:      $\{x \in \mathbb{R}: 1 \le x \le 4\}$

**18**   Function:    $f(x) = \dfrac{1}{x}$,            Domain:      $\{x \in \mathbb{R}: 0 < x \le 1\}$

**19**   Function:    $f(x) = x^2 - 1$,          Domain:      $\mathbb{R}$

**20**   Function:    $f(x) = x^2 + 4$,          Domain:      $\mathbb{R}$

**21**   Function:    $f(x) = \dfrac{1}{x-1}$,        Domain:      $\{x \in \mathbb{R}: x \ne 1\}$

**22**   Function:    $f(x) = \dfrac{x+1}{x-1}$,       Domain:      $\{x \in \mathbb{R}: x \ne 1\}$

For questions **23** to **28** state whether the function is one-to-one or many-to-one for the stated domain.

**23** $f(x) = x$, domain: $\mathbb{R}$

**24** $f(x) = x^2$, domain: $\{x \in \mathbb{R}: 0 \le x \le 3\}$

**25** $f(x) = x^2$, domain: $\{x \in \mathbb{R}: -3 \le x \le 3\}$

**26** $f(x) = x^2$, domain: $\mathbb{R}$

**27** $f(x) = \sqrt{x}$, domain: $\{x \in \mathbb{R}: 1 \le x \le 4\}$

**28** $f(x) = \sqrt{x}$, for the natural domain of the function

State the natural domain and corresponding range for each of the following.

**29** $f(x) = 2x + 3$

**30** $f(x) = x^2$

**31** $f(x) = \sqrt{x}$

**32** $f(x) = \sqrt{x - 3}$

**33** $f(x) = \sqrt{x + 3}$

**34** $f(x) = 5 + \sqrt{x - 3}$

**35** $f(x) = \dfrac{1}{x - 3}$

**36** $f(x) = \dfrac{1}{\sqrt{x - 3}}$

# Miscellaneous exercise three

**This miscellaneous exercise may include questions involving the work of this chapter, the work of any previous chapters, and the ideas mentioned in the Preliminary work section at the beginning of the book.**

**1** Solve    **a**   $\dfrac{2x - 1}{3} = \dfrac{3x + 2}{5}$       **b**   $\dfrac{3x - 1}{2} + 7 = \dfrac{2x + 7}{3}$

**2** Find the range of the function $f(x) = 3 - 2x$ for the domain $\{1, 2, 3, 4\}$.

**3** Graph 1 below shows the graph of $f(x) = x + 1$ for the domain $-2 \le x \le 3$.

Graph 2 shows the graph of $f(x) = x + 1$ for the domain $\{-2, -1, 0, 1, 2, 3\}$.

State the range of the function for each domain.

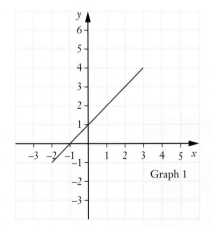

Graph 1

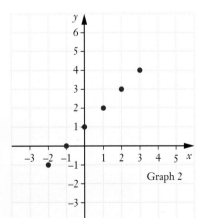

Graph 2

**4** Expand and then simplify each of the following:

   **a**  $(a + b)^2$      **b**  $(a + b)^3$      **c**  $(a + 2b)^3$      **d**  $(a - 2b)^3$

**5** For each of the following diagrams, state whether the relationship shown is a function or not and, for those that are functions, state whether the function is one-to-one or many-to one.

   **a**  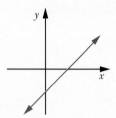    **b**  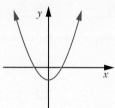    **c**  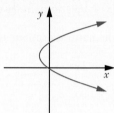

   **d**  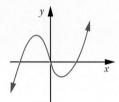    **e**  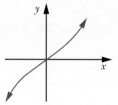    **f**  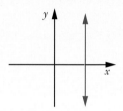

**6** The isosceles triangle ABC has AB = AC = 10 cm and BC = 12 cm. Three circles are drawn, one with centre A, radius 4 cm, another with centre B, radius 6 cm, and a third with centre C, radius 6 cm. Find the area of that part of triangle ABC not lying in any of the circles.

**7** Three ships A, B and C are such that B is 4.8 km from A on a bearing 115° and C is 5.7 km from A on a bearing 203°.

How far and on what bearing is B from C?

**8** The diagram on the right shows the sketch made by a surveyor after taking measurements for a block of land. Find the area of the block.

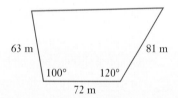

**9** A pig farmer makes a feeding trough for his pigs by cutting a cylindrical metal drum (see diagram). If the drum is of length 120 cm and radius 30 cm and the cut is made 10 cm from the axis of the drum find the capacity of the trough correct to the nearest litre. (Ignore the thickness of the metal.)

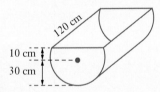

ISBN 9780170390330

# 4.

# Linear functions

- Revision of straight line graphs
- Lines parallel to the axes
- Direct proportion and straight line graphs
- Further considerations
- Determining the equation of a straight line
- Calculator and internet routines
- Parallel and perpendicular lines
- Miscellaneous exercise four

## Situation

Three electricians, Sparky, Flash and Voltman, have different ways of calculating a customer's bill.

- Sparky charges a standard rate per hour and has no other charges.
- Flash has a fixed, or 'standing' charge and then charges a certain amount per hour on top of that.
- Voltman has a higher standing charge than Flash but then charges less per hour.

These three methods are shown graphed below:

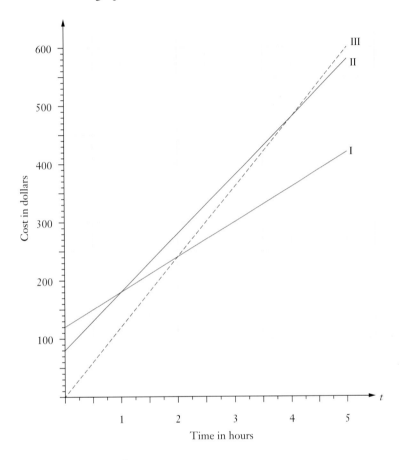

- Which line, I, II or III, corresponds to

  **a**  Sparky?                **b**  Flash?                **c**  Voltman?

- Ignoring the standing charges, who charges most per hour? What feature of the graph shows this?
- With the charge being $C and the time being $t$ hours the equation of line I is:

$$C = 60t + 120.$$

  Determine the equations of lines II and III.

- If you were considering using one of the three electricians for a job and wanted to keep the cost to a minimum which of the three could you dismiss from your considerations?

It is likely that you have already encountered straight line graphs in your mathematical studies of earlier years. Indeed the *Preliminary work* at the beginning of this text included brief mention of the equation

$$y = mx + c.$$

Reading through the next few pages, and then working through **Exercise 4A** which follows, should revise these ideas and provide further practice.

# Revision of straight line graphs

x- and y-intercepts

For each of the three electricians in the situation on the previous page, the amount they charged the customer, $C, and the time of the job, $t$ hours, were ***linearly related***. I.e. graphing paired values of the two variables $t$ and $C$ gave a *straight line* in each case.

Two important features of straight line graphs are:

- The steepness or gradient of the line.
  In the situation on the previous page this feature indicated the hourly rate charged by each electrician (neglecting the standing charge).

and
- The point where the line cuts the vertical axis.
  In the situation on the previous page this feature indicated the standing charge for each electrician.

Consider these two features for the graph shown on the right:

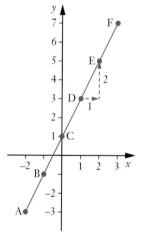

- Each time we move to the right 1 unit the line goes up 2 units. We say that the line has a *gradient*, or *slope*, of 2 units. (If a straight line goes *down* for each unit we move right we say it has a *negative* gradient.)

- The line cuts the vertical axis at the point with coordinates (0, 1).

Looking at points A, B, C, D, E and F, all lying on the line, the following table can be created:

|  | A | B | C | D | E | F |
|---|---|---|---|---|---|---|
| **x coordinate** | −2 | −1 | 0 | 1 | 2 | 3 |
| **y coordinate** | −3 | −1 | 1 | 3 | 5 | 7 |

Notice that as the $x$ values in the table increase by 1, the $y$ values increase by 2, as we would expect for a table of values for a line with a gradient of 2. If for each unit increase in $x$ the $y$ values did not increase by a constant amount, the points would not lie in a straight line.

From the table we see that the rule or equation governing these pairs of numbers is

$$y = 2x + 1.$$

> If a straight line cuts the $y$-axis at $(0, c)$ and has a gradient of $m$ then its equation is:
>
> $$y = mx + c$$

For example:

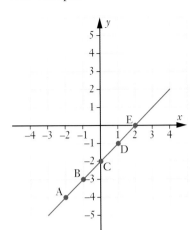

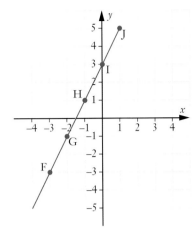

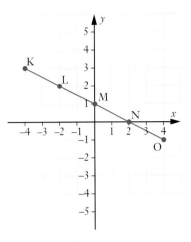

Gradient of line = 1

Cuts $y$-axis at $(0, -2)$

Rule: $y = 1x - 2$

Which agrees with the following table:

Gradient of line = 2

Cuts $y$-axis at $(0, 3)$

Rule: $y = 2x + 3$

Which agrees with the following table:

Gradient of line = $-0.5$

Cuts $y$-axis at $(0, 1)$

Rule: $y = -0.5x + 1$

Which agrees with the following table:

|   | A | B | C | D | E |
|---|---|---|---|---|---|
| $x$ | -2 | -1 | 0 | 1 | 2 |
| $y$ | -4 | -3 | -2 | -1 | 0 |

|   | F | G | H | I | J |
|---|---|---|---|---|---|
| $x$ | -3 | -2 | -1 | 0 | 1 |
| $y$ | -3 | -1 | 1 | 3 | 5 |

|   | K | L | M | N | O |
|---|---|---|---|---|---|
| $x$ | -4 | -2 | 0 | 2 | 4 |
| $y$ | 3 | 2 | 1 | 0 | -1 |

## EXAMPLE 1

State the equation of the straight line that cuts the $y$-axis at the point $(0, 5)$ and has a gradient of 7.

### Solution

A line with gradient $m$ and cutting the $y$-axis at $(0, c)$ has equation $y = mx + c$.
Thus the given line has equation $y = 7x + 5$.

## EXAMPLE 2

A straight line has equation $y = 3x - 5$. Determine its gradient and the coordinates of the point where it cuts the $y$-axis.

### Solution

A line with equation $y = mx + c$ has gradient $m$ and cuts the $y$-axis at $(0, c)$.
Thus the given line has gradient 3 and cuts the $y$-axis at the point $(0, -5)$.

If the equation of a straight line is not presented in the form $y = mx + c$ some initial rearrangement may be made to present it in this form, as the next example shows.

## EXAMPLE 3

A straight line has equation $3x + 5y = 20$. Determine its gradient and the coordinates of the point where it cuts the $y$-axis.

**Solution**

Given: $\qquad\qquad\qquad\qquad\qquad\qquad\qquad 3x + 5y = 20$

Subtract $3x$ from each side to isolate $5y$: $\qquad 5y = -3x + 20$

Divide each side by 5 to isolate $y$: $\qquad\qquad y = -0.6x + 4$

This is now of the form $y = mx + c$.

Thus the given line has gradient $-0.6$ and cuts the $y$-axis at the point $(0, 4)$.

The equation of a line is like the 'membership ticket' for points lying on the line:

The coordinates of every point lying on a straight line will 'fit' the equation of the line and every point not lying on the line will not fit the equation.

## EXAMPLE 4

Determine whether or not the point $(-2, -8)$ lies on the line $y = -3x - 14$.

**Solution**

If $(-2, -8)$ lies on the given line then substituting the $x$-coordinate, $-2$, into the equation should give the $y$-coordinate, $-8$.

If $\quad x = -2 \quad$ then $\quad y = -3(-2) - 14$

$\qquad\qquad\qquad\qquad\quad = 6 - 14$

$\qquad\qquad\qquad\qquad\quad = -8$

Thus the point $(-2, -8)$ *does* lie on the line $y = -3x - 14$.

# Lines parallel to the axes

Lines parallel to the $x$-axis have zero gradient.

They will have equations of the form $\quad y = 0x + c,$

$\qquad\qquad\qquad$ i.e. $\quad y = c.$

Hence the graph on the right shows the lines

$\qquad\qquad\qquad\qquad y = 6,$

$\qquad\qquad\qquad\qquad y = 3$

and $\quad y = -2.$

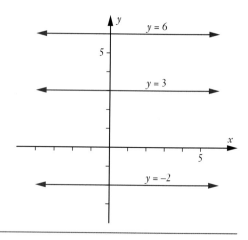

Lines parallel to the $y$-axis have an undefined gradient – we cannot find the vertical rise in the line for each horizontal unit increased because the line rises vertically for zero increase horizontally! Hence we should not expect the rules for vertical lines to be of the form $y = mx + c$ because the gradient, $m$, is undefined. Indeed straight lines parallel to the $y$-axis are the only straight lines having rules that are *not* of the form $y = mx + c$.

Lines parallel to the $y$-axis have rules of the form    $x = c$.

The graph on the right shows the vertical lines:    $x = -2$,

$$x = 3$$

and    $x = 5$.

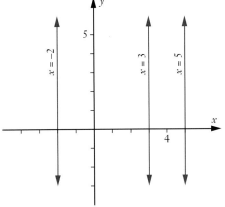

Even though these vertical lines have rules of a different form, points lying on each line must still 'obey' the rule. For example, for a point to lie on $x = 3$ the point must have an $x$-coordinate equal to 3.

# Direct proportion and straight line graphs

As we were reminded in the *Preliminary work* section at the beginning of this text, if $x$ and $y$ are **directly proportional** (also called **direct variation**) then

- if $x$ is doubled, $y$ is doubled
- if $x$ is trebled, $y$ is trebled, etc.

If $x$ and $y$ are directly proportional, and if when $x = 1$, $y = k$, the following table would result:

| $x$ | 1 | 2 | 3 | 4 | 5 | 6 |
|---|---|---|---|---|---|---|
| $y$ | $k$ | $2k$ | $3k$ | $4k$ | $5k$ | $6k$ |
| Difference | | $k$ | $k$ | $k$ | $k$ | $k$ |

As the $x$ values increase by 1 the $y$ values show a constant difference pattern of $k$.

Thus the rule will be of the form    $y = kx + c$.

From the values in the table    $c = 0$.

Hence, for direct proportion, the rule relating $x$ and $y$ is $y = kx$, i.e. a straight line of gradient $k$ and passing through the point $(0, 0)$.

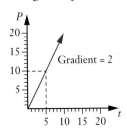

Direct proportion

Rule: $P = 2t$

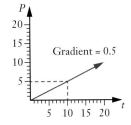

Direct proportion

Rule: $P = 0.5t$

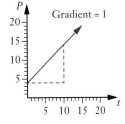

Not direct proportion

Rule: $P = t + 4$

## Exercise 4A

**1** For each of the lines A to L shown below, write down

   **a**   the coordinates of the point where the line cuts the $y$-axis

   **b**   the gradient of the line

   **c**   the equation of the line

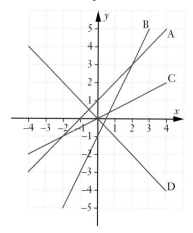 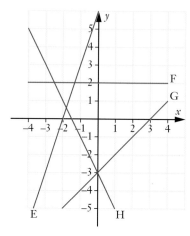 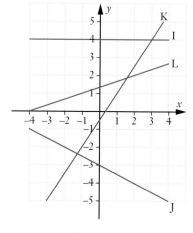

**2** For each of the following tables determine whether the paired values $(x, y)$ would, if graphed, give points that lie in a straight line. For those that do determine the equation of the straight line. (You should be able to do this question from the tables and should not need to plot the points.)

**a**

| $x$ | 1 | 2 | 3 | 4 | 5 | 6 |
|---|---|---|---|---|---|---|
| $y$ | 7 | 9 | 11 | 13 | 15 | 17 |

**b**

| $x$ | 1 | 2 | 3 | 4 | 5 | 6 |
|---|---|---|---|---|---|---|
| $y$ | −2 | 3 | 8 | 13 | 18 | 23 |

**c**

| $x$ | 1 | 2 | 3 | 4 | 5 | 6 |
|---|---|---|---|---|---|---|
| $y$ | 0 | 1 | 3 | 6 | 10 | 15 |

**d**

| $x$ | 1 | 2 | 3 | 4 | 5 | 6 |
|---|---|---|---|---|---|---|
| $y$ | −3 | −2 | −1 | 0 | 1 | 2 |

**e**

| $x$ | 0 | 1 | 2 | 3 | 4 | 5 |
|---|---|---|---|---|---|---|
| $y$ | 10 | 8 | 6 | 4 | 2 | 0 |

**f**

| $x$ | 0 | 1 | 2 | 3 | 4 | 5 |
|---|---|---|---|---|---|---|
| $y$ | 5 | 5 | 5 | 5 | 5 | 5 |

**g**

| $x$ | 3 | 4 | 1 | 6 | 2 | 5 |
|---|---|---|---|---|---|---|
| $y$ | 9 | 4 | 16 | −9 | 13 | −2 |

**h**

| $x$ | 4 | 2 | 1 | 6 | 3 | 5 |
|---|---|---|---|---|---|---|
| $y$ | 7 | −3 | −8 | 17 | 2 | 12 |

**3** Copy and complete the following table.

| Equation | Gradient | $y$-axis intercept |
|---|---|---|
| $y = 2x + 3$ | ? | (0, ?) |
| $y = 3x + 4$ | ? | (0, ?) |
| $y = -2x - 7$ | ? | (0, ?) |
| $y = 6x + 3$ | ? | (0, ?) |

**4** Write down the equation of the straight line with gradient 4 and cutting the $y$-axis at $(0, 6)$.

**5** Write down the equation of the straight line cutting the $y$-axis at $(0, -5)$ and with gradient $-1$.

**6** Suppose that a particular 'family' of straight lines are all those with a gradient of 2. Which of the following straight lines are in this family?

Line A:   Equation   $y = 3x + 2$          Line B:   Equation   $y = 2x - 3$

Line C:   Equation   $y = 2$               Line D:   Equation   $y = 2x$

Line E:   Equation   $y = 5 + 2x$          Line F:   Equation   $2y = 4x + 7$

Line G:   Equation   $y - 2x = 5$          Line H:   Equation   $3y + 6x = 5$

**7** Suppose that a particular 'family' of straight lines are all those passing through the point $(0, 6)$. Which of the following straight lines are in this family?

Line A:   Equation   $y = 5x + 6$          Line B:   Equation   $y = 6x + 5$

Line C:   Equation   $y = 6x$              Line D:   Equation   $y = 6$

Line E:   Equation   $y = 6 + x$           Line F:   Equation   $y + 6 = x$

Line G:   Equation   $2y = -x + 12$        Line H:   Equation   $x + y = 6$

**8** Write down the equation of the straight line with gradient $-4$ and cutting the $y$-axis at $(0, -3)$. Does this line pass through the point $(-1, 1)$?

**9** Write down the equation of the straight line cutting the $y$-axis at $(0, -3)$ and with gradient 2. Which of the following points lie on this line?

A$(5, 7)$,        B$(-3, -1)$,        C$(-0.5, -4)$,        D$(2.5, 2)$,        E$(-2, -1)$.

**10** Copy and complete the following table.

| Equation | Written as $y = mx + c$ | Gradient | $y$-axis intercept |
|---|---|---|---|
| $2y = 4x - 5$ | | ? | $(0, ?)$ |
| $4y = 3x + 7$ | | ? | $(0, ?)$ |
| $3y - 2x = 6$ | | ? | $(0, ?)$ |
| $4x + 3y - 6 = 0$ | | ? | $(0, ?)$ |
| $3x + 5y = 8$ | | ? | $(0, ?)$ |

**11** Points A$(3, a)$, B$(5, b)$ and C$(c, -9)$ all lie on the line $y = 7x + 5$. Find $a$, $b$ and $c$.

**12** The points D, E, F, G, H and I, whose coordinates are given below, all lie on the line $y = dx - 5$.

D$(4, -3)$,        E$(8, e)$,        F$(-2, f)$,        G$(13, g)$,        H$(h, -4.5)$,        I$(i, -7.5)$.

Determine the values of $d, e, f, g, h$ and $i$.

**13** For each of the following graphs, state whether $P$ and $t$ vary directly with each other (i.e. are directly proportional to each other) or not and, for those cases when direct proportion is involved, find the rule for the relationship.

**a**
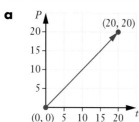

**b**
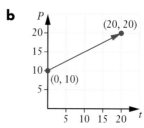

**c**
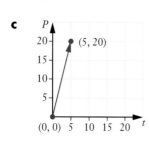

**d**
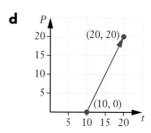

**e**
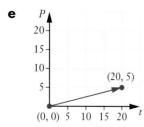

**f**
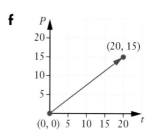

**g**
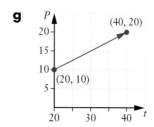

**h**
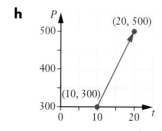

# Further considerations

Our understanding of coordinates, gradients and the Pythagorean theorem allows us to:

- determine the coordinates of the mid-point of a line joining two points
- determine the gradient of a line joining two points
- determine the distance between two points.

ISBN 9780170390330

Consider moving from the point A(–2, 3) to B(6, 9). This involves moving *right* 8 units and *up* 6 units, as shown below left.

If, from point A, we only wanted to move half way towards point B, we would move just 4 units right and 3 units up. This would take us to the point (2, 6), as shown below right.

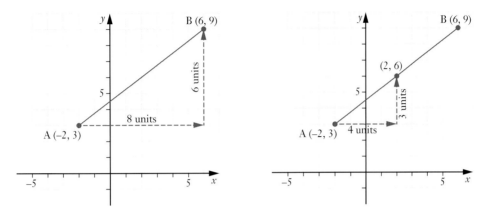

Note that as you may have expected, the *x*-coordinate of the midpoint of the line joining points A and B is the mean of the *x*-coordinates of the two points, and the *y*-coordinate of the mid-point is the mean of the *y*-coordinates of the two points. Hence, if we want to avoid having to plot the points on a graph the following result can be used directly:

The coordinates of the **midpoint of the line joining point A to point B** will be:

(the mean of the two *x*-coordinates, the mean of the two *y*-coordinates).

Thus if A has coordinates $(x_1, y_1)$ and B has coordinates $(x_2, y_2)$ then the coordinates of the midpoint will be given by:

$$\left( \frac{x_1 + x_2}{2}, \frac{y_1 + y_2}{2} \right)$$

Now consider the points A(–6, 4) and B(4, –2) shown graphed below left. To move from A to B we move *right* 10 units and *down* 6 units, as shown below right.

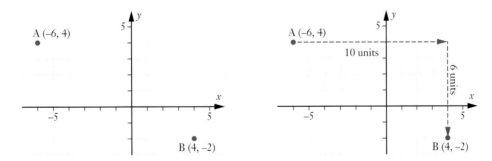

Hence in moving right one unit we move down 0.6 units. Thus the gradient of the line through A and B is –0.6.

If we want to avoid having to plot the points on a graph the following result can be used directly:

If a line passes through two points, A and B, then the **gradient of the line** is:

$$\frac{\text{the change in the } y\text{-coordinate in going from A to B}}{\text{the change in the } x\text{-coordinate in going from A to B}}.$$

Thus if A has coordinates $(x_1, y_1)$ and B has coordinates $(x_2, y_2)$ then:

Gradient of the straight line through $A(x_1, y_1)$ and $B(x_2, y_2) = \dfrac{y_2 - y_1}{x_2 - x_1}$.

Note:

- In the previous formula $\dfrac{y_1 - y_2}{x_1 - x_2}$ would also give the correct answer but $\dfrac{y_1 - y_2}{x_2 - x_1}$ and $\dfrac{y_2 - y_1}{x_1 - x_2}$ would not.

To find the length of the straight line joining A(–6, 4) to B(4, –2) we use the Pythagorean theorem:

$$AB^2 = 10^2 + 6^2$$
$$= 100 + 36$$
$$= 136$$
$$AB = \sqrt{136}$$
$$= 11.7 \text{ units (to 1 decimal place)}$$

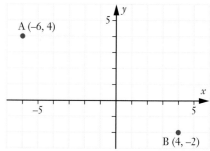

The length of the line joining A(–6, 4) to B(4, –2) is 11.7 units, to one decimal place.

Again, if we want to apply a formula and not plot the points:

The **length of the line** joining point A, $(x_1, y_1)$, to point B, $(x_2, y_2)$, is:

$$\sqrt{(y_2 - y_1)^2 + (x_2 - x_1)^2}.$$

I.e. $\sqrt{(\text{change in the } y\text{-coordinate})^2 + (\text{change in the } x\text{-coordinate})^2}.$

## EXAMPLE 5

Find the coordinates of the midpoint of AB given that point A has coordinates (–7, 19) and point B has coordinates (5, –3).

**Solution**

Coordinates of midpoint $= \left(\dfrac{-7 + 5}{2}, \dfrac{19 + (-3)}{2}\right)$, i.e. (–1, 8).

The midpoint of the line AB has coordinates (–1, 8).

ISBN 9780170390330

**EXAMPLE 6**

Find the gradient of the straight line through C(3, –5) and D(6, 4).

**Solution**

Gradient $= \dfrac{(4)-(-5)}{(6)-(3)}$

$= \dfrac{9}{3}$

$= 3$

The straight line through C and D has a gradient of 3.

**EXAMPLE 7**

Find the length of the straight line joining point E(1, –7) to point F(13, –2).

**Solution**

$EF^2 = (-7--2)^2 + (1-13)^2$

$= (-5)^2 + (-12)^2$

$= 169$

$EF = \sqrt{169}$

$= 13$ units

The length of the line joining E(1, –7) to F(13, –2) is 13 units.

Your calculator, and various internet sites, may have programmed routines that allow

- the coordinates of the midpoint of a line joining two points,
- the gradient of the straight line through two points,

and
- the distance between two points,

to be determined.

**Note**

Such routines can be useful but make sure that you understand the underlying ideas as shown in the previous examples and can apply them without the assistance of calculator and internet programs if required.

## Exercise 4B

**1** Calculate the coordinates of the midpoint of the straight line joining each of the following pairs of points.

**a** (4, 6) and (10, 12)     **b** (6, 7) and (4, 13)     **c** (4, 5) and (2, 5)

**d** (–6, 7) and (2, –5)     **e** (0, 5) and (–4, 2)     **f** (5, 3) and (19, –1)

**g** (6, –2) and (10, –9)     **h** (–5, 12) and (5, 3)     **i** (–6, 8) and (8, –6)

**2** Find the gradient of the straight line through each of the following pairs of points.

**a** (4, 6) and (2, 2)     **b** (6, 7) and (7, 3)     **c** (4, 5) and (2, 1)

**d** (6, 7) and (2, 5)     **e** (5, 3) and (1, 4)     **f** (3, 3) and (4, 2)

**g** (4, 3) and (2, 7)     **h** (5, 2) and (3, –3)     **i** (4, 2) and (–2, –1)

**3** Calculate the length of the straight line joining each of the following pairs of points.

    **a**   (4, 6) and (7, 10)         **b**   (6, 7) and (3, 11)         **c**   (4, 5) and (–8, 10)

    **d**   (6, 1) and (–1, 25)         **e**   (5, –3) and (–3, 12)        **f**   (–6, 8) and (0, 0)

    **g**   (4, 3) and (11, 4)         **h**   (5, 2) and (–2, 5)         **i**   (9, 9) and (3, 4)

**4** Points C and D have coordinates (3, 6) and (4, 8) respectively. Find

    **a**   the gradient of the straight line joining C and D

    **b**   the length of the straight line joining C and D

    **c**   the coordinates of the midpoint of the line CD.

**5** Points E and F have coordinates (–1, 1) and (4, 9) respectively. Find

    **a**   the gradient of the straight line joining E and F

    **b**   the length of the straight line joining E and F

    **c**   the coordinates of the midpoint of the line EF.

**6** The length of the straight line joining point A(1, 4) to point B(7, $c$) is 10 units. Find the two possible values of $c$.

**7** The diagram on the right shows the location of point A on the mainland and B and C on islands. Each unit shown on the graph is 1 km.

    Calculate the distance from

    **a**   A to B

    **b**   A to C

    **c**   B to C

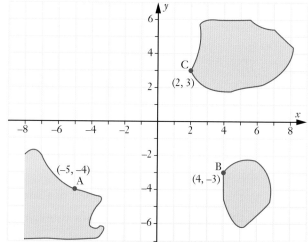

**8** The diagram below shows a simplified model of the three stages of a proposed mountain climb.

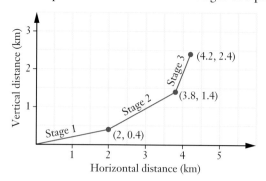

    According to this linear model, what is the gradient of each stage?

ISBN 9780170390330

# Determining the equation of a straight line

Equations of lines

As we have already seen, knowing the gradient of a straight line and the coordinates of the point where the line cuts the vertical axis we can determine the equation of the straight line. Similarly we have seen that we can determine the equation of a straight line given the graph of the line or the table of values for the line. Two other common situations that may occur and for which the given information allows us to determine the equation of the line are as follows:

- Given the gradient of the line and the coordinates of *any* point on the line, not necessarily the vertical intercept. (See **Example 8**.)
- Given the coordinates of any two points that lie on the line. (See **Example 9**.)

## EXAMPLE 8 (Given the gradient and one other point on the line.)

Find the equation of the straight line through the point $(4, -3)$ and with a gradient of $-2$.

### Solution

A straight line of gradient $m$ has an equation of the form $\qquad y = mx + c$.

Thus the given line will have an equation of the form $\qquad y = -2x + c$.

The line passes through the point $(4, -3)$.

Thus the values $x = 4$ and $y = -3$ must 'fit' the equation, i.e. $\qquad (-3) = -2(4) + c$,

giving $\qquad c = 5$.

Thus the given line has equation $y = -2x + 5$.

## EXAMPLE 9 (Given two points that lie on the line.)

Find the equation of the straight line through the points $(-2, 8)$ and $(4, -1)$.

### Solution

First we determine the gradient of the line, either by reasoning:

Starting with the point with the lower $x$-coordinate, $(-2, 8)$, and moving to the other point, $(4, -1)$,

we travel across 6 units and down 9 units. Thus in moving across 1 unit we travel *down* $\dfrac{9}{6}$ units,

i.e. $\dfrac{3}{2}$ units. The gradient of the line is $-\dfrac{3}{2}$.

Or by use of $\dfrac{y_2 - y_1}{x_2 - x_1}$: $\qquad\qquad$ Gradient $= \dfrac{8 - (-1)}{-2 - 4}$

$$= -1.5.$$

Thus the given line will have an equation of the form $\qquad y = -1.5x + c$.

The line passes through the point $(4, -1)$. $\qquad$ Thus $\quad -1 = -1.5(4) + c$,

giving $\quad c = 5$.

Thus the given line has equation $y = -1.5x + 5$.

(The reader should confirm that using the point $(-2, 8)$ and saying that the values $x = -2$ and $y = 8$ must 'fit' the equation also gives $c = 5$.)

ISBN 9780170390330

EXAMPLE 10 **(Given information that allows us to determine two points that lie on the line and then proceed as in the previous example.)**

The line $y = 2x - 6$ cuts the $x$-axis at the point A.

The line $y = 5 - 7x$ passes through the point B, coordinates $(-1, k)$.

Find the equation of the straight line through points A and B.

**Solution**

Any point on the $x$-axis has a $y$-coordinate of zero. Thus at point A:   $0 = 2x - 6$,

$$\therefore \quad x = 3.$$

Point A has coordinates $(3, 0)$.

Point B $(-1, k)$ lies on the line $y = 5 - 7x$.         Thus   $k = 5 - 7(-1)$,

$$\therefore \quad k = 12.$$

Point B has coordinates $(-1, 12)$.

The line through points A$(3, 0)$ and B$(-1, 12)$ has gradient     $\dfrac{12 - 0}{-1 - 3} = -3.$

Thus the required equation is of the form         $y = -3x + c.$

But the point $(3, 0)$ lies on this line, hence:         $0 = -3(3) + c,$

$$\therefore \quad c = 9.$$

The required equation is $y = -3x + 9$.

# Calculator and internet routines

As with determining the coordinates of the **midpoint of a line** joining two points, the **distance between two points**, and the **gradient of the straight line** through two points, your calculator and some internet sites may also have programmed routines that allow the equation of a line to be determined simply by inputting the coordinates of two points on the line, or inputting the gradient and the coordinates of just one point on the line. Again such routines can be useful but make sure that you understand the underlying ideas and can apply them without the assistance of such programs if required.

### Exercise 4C

**1** Write the equations of each of the lines A to J shown in the graphs below.

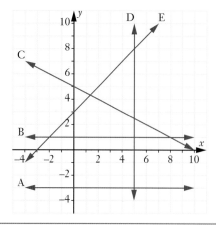
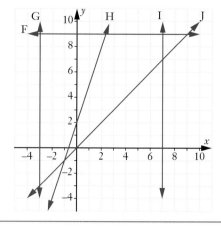

**2** What is the equation of the *x*-axis?

**3** What is the equation of the *y*-axis?

**4** Write down the equation of the straight line with gradient 3 and cutting the *y*-axis at (0, 4). Does this line pass through the point (–1, 1)?

**5** Write down the equation of the straight line cutting the *y*-axis at (0, 2) and with gradient 0.5. Which of the following points lie on this line ?

A(2, 1),     B(2, 0),     C(4, 2),     D(–6, –1),     E(4, 4).

**6** Find the equation of each of the following straight lines.
  **a**  Gradient 1, through (3, 5).
  **b**  Gradient –1, through (6, –1).
  **c**  Gradient –2, through (3, 2).
  **d**  Gradient 5, through (–2, –2).
  **e**  Gradient 0.5, through (8, 9).
  **f**  Gradient –0.5, through (–3, 0).
  **g**  Gradient 1.5, through (9, 2).
  **h**  Gradient $-\frac{1}{3}$, through (7, –1).

**7** Find the equation of each of the following straight lines.
  **a**  Through (2, 5) and (6, 9).
  **b**  Through (0, –1) and (2, –9).
  **c**  Through (14, 1) and (16, –5).
  **d**  Through (1, 1) and (2, 3).
  **e**  Through (1, 2) and (13, 6).
  **f**  Through (3, –2) and (–1, 6).
  **g**  Through (3, 9) and (0, 4).
  **h**  Through (0, 5) and (2, –5).

**8** Find the equation of the straight line passing through (1, 1) and (4, 7). Determine which of the points listed below lie on this line.

A(7, 15),     B(7, 13),     C(2, 2),     D(–1, 3),     E(6, 11).

**9** Find the equation of the straight line with gradient 0.5 and passing through the point (3, 4). Given that each of the points listed below lie on this line, determine the values of *f*, *g*, *h*, *i* and *j*.

F(9, *f*),     G(–9, *g*),     H(*h*, 9),     I(*i*, 1.5),     J(3.8, *j*).

**10** Find the coordinates of the point where the line $2y = x - 4$ cuts the *x*-axis. Find the equation of the straight line through this point and (–1, 10).

**11** Find the coordinates of the point where the line $2y = -x + 6$ cuts the *x*-axis. Find the equation of the straight line through this point and the point (8, 8).

**12** If we plot degrees Celsius, (°C), on the *x*-axis and degrees Fahrenheit, (°F), on the *y*-axis, the graph for converting from one scale to the other is a straight line. Given that 100°C is the same as 212°F and 50°C is the same as 122°F find the equation of the line in the form $F = mC + b$, where *m* and *b* are constants.

Convert the following to °F.
  **a**  55°C
  **b**  125°C
  **c**  –10°C

Convert the following to °C.
  **d**  59°F
  **e**  86°F
  **f**  –40°F

**13** If we plot the 'Number of metered units', $N$, on the $x$-axis and the 'Amount to be paid', $A$, on the $y$-axis then the graph for calculating a telephone bill is a straight line.

If the bill for 100 units is \$64 and for 175 units is \$82, determine the equation of this line in the form $A = mN + c$, where $m$ and $c$ are constants.

**14** The diagram on the right shows the proposed layout of a small airfield. The diagram shows the main runway, the approach lights, the warning lights and the administration building.

The second diagram shows the proposal as a graph with lengths in metres and the admin building as the origin.

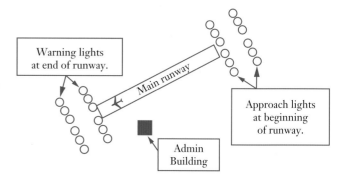

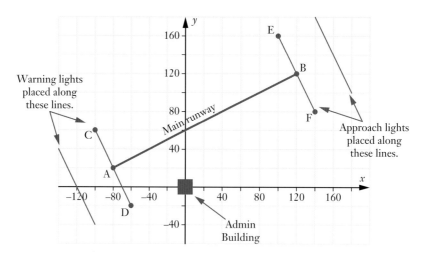

Find:

**a** the coordinates of the points A, B, C, D, E and F (all divisible by 20)

**b** the length AB

**c** the equation of the straight line through A and B

**d** the equation of the straight line through C and D

**e** the equation of the straight line through E and F.

**15** At 8 a.m. one morning an industrial fuel tank contains 4000 litres of fuel.
Fuel is being withdrawn from the tank at a constant rate of 0.25 litres per minute.

Writing $A$ litres for the amount of fuel in the tank at time $t$ hours past 8 a.m. find $A$ when $t = 2$ and $t$ when $A = 3850$.

Express the relationship between $A$ and $t$ in the form $A = mt + c$.

ISBN 9780170390330

**16** Susie Fuse, an electrician, charges her customers a standard call out fee plus a certain amount per hour. For a job that takes her 3 hours she charges \$445 and for a job that takes her 4.5 hours she charges \$625.

Write her method of charging in the form $C = mT + c$, where \$C$ is the cost to the customer for a job that takes $T$ hours and $m$ and $c$ are constants.

**17** A linear relationship exists between the profit, \$P$, that the organisers of a concert make, and $N$, the number of tickets they sell. With $P$ plotted on the vertical, $y$, axis and $N$ on the horizontal, $x$, axis the line of this relationship passes through the points (900, 400) and (1100, 1300). Find the equation of this line in the form

$$P = mN + c,$$

where $m$ and $c$ are constants.

**a** What will be the profit when 1500 tickets are sold?

**b** If the concert hall has a maximum capacity of 2500 what profit will the organisers make if they give away 150 complimentary tickets and sell all the rest?

**c** What is the least number of tickets the organisers could sell and still not make a loss?

**18** The owner of a computer shop calculates that his weekly profit from computer sales is linearly related to the number of computers sold that week.

If he sells 10 computers in a week his total profit is \$560.

If he only sells 5 computers in the week he makes a profit of \$10.

The rule relating his total profit for the week, \$P$, to the number of computers sold, $x$, is given by:

Total profit in dollars $= mx - c$,

\$c$ being the fixed weekly cost of running the shop.

**a** Calculate $m$ and $c$.

**b** What is his weekly profit from computer sales in a week that he sells 20 computers?

**19** When a particular spring, of unstretched length $L_0$ metres, has a mass of $M$ kg suspended from one end its new length, $L$ metres, is given by:

$$L = kM + L_0 \text{ where } k \text{ is a constant.}$$

A graph of $M$ plotted on the horizontal axis and $L$ on the vertical axis passes through the points (2, 0.85) and (3, 1.05).

Calculate $k$ and $L_0$ and hence determine how much the spring is extended **beyond its natural length** when a mass of 250 g is suspended from it.

Parallel and perpendicular lines

# Parallel and perpendicular lines

Two lines that are parallel must have the same gradient but what is the relationship between the gradients of two perpendicular lines?

Each of the following diagrams show two perpendicular lines. Can you notice any pattern in the gradients of two lines that are perpendicular?

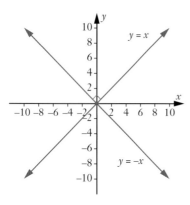

Gradients of 1 and –1.

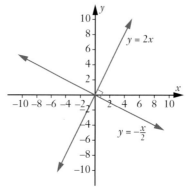

Gradients of 2 and $-\frac{1}{2}$.

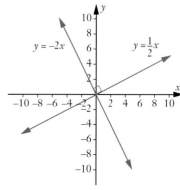

Gradients of –2 and $\frac{1}{2}$.

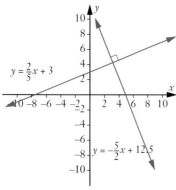

Gradients of $\frac{2}{5}$ and $-\frac{5}{2}$.

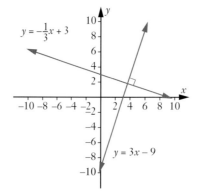

Gradients of $-\frac{1}{3}$ and 3.

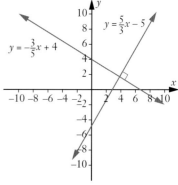

Gradients of $-\frac{3}{5}$ and $\frac{5}{3}$.

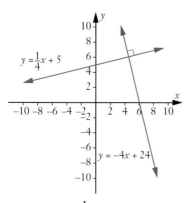

Gradients of $\frac{1}{4}$ and –4.

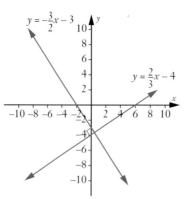

Gradients of $-\frac{3}{2}$ and $\frac{2}{3}$.

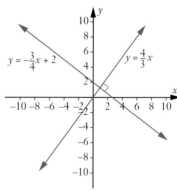

Gradients of $-\frac{3}{4}$ and $\frac{4}{3}$.

ISBN 9780170390330

Did you notice that:

If the gradient of one line is $m$, a perpendicular line has gradient $-\dfrac{1}{m}$?

This can be summarised as follows:

> The gradients of two perpendicular lines have a product of $-1$.

Note:   The only situation involving perpendicular lines where this rule does not apply is that of a horizontal line, e.g. $y = 3$ (zero gradient), and a vertical line, e.g. $x = 4$ (infinite gradient). In such cases the product of the gradients does not equal $-1$ but the lines are perpendicular.

## Exercise 4D

**1** The eleven lines whose equations are given below contain five pairs of parallel lines and one that is not parallel to any of the others.

List the five pairs. (i.e. A and E, ? and ?, … ).

A:   $y = 2x + 3$      B:   $y = 3x + 4$      C:   $y = 5x + 3$      D:   $y = \dfrac{1}{2}x + 3$

E:   $y = 2x - 1$      F:   $y = 5 - \dfrac{1}{2}x$      G:   $y + 5x = 1$      H:   $y - 5x = 4$

I:   $y = 1 - 5x$      J:   $y = 3x - 2$      K:   $2y + x = 6$

**2** Find the equation of the straight line through the point $(-1, -7)$ and parallel to the line $y = 2x + 3$.

**3** The eleven lines whose equations are given below contain five pairs of perpendicular lines and one that is not perpendicular to any of the others.

List the five pairs.

A:   $y = -2x + 3$      B:   $y = 3x$      C:   $y = x + 3$      D:   $y = \dfrac{1}{2}x + 1$

E:   $y = -x + 1$      F:   $y = 3$      G:   $3y + x = 3$      H:   $2y = 3x + 2$

I:   $2y + 3x = 8$      J:   $3y = 2x - 9$      K:   $x = -2$

**4** Find the equation of the straight line that is perpendicular to $y = 2x + 3$ and passes through the point $(-4, 7)$.

**5** Find the equation of the straight line that is perpendicular to $3y = 5 - x$ and passes through the point $(-1, 2)$.

**6 a** The lines $y = x - 3$ and $y = 3x - 7$ intersect at the point B.

Using your calculator, or otherwise, find the coordinates of point B.

**b** Find the equation of the line that is perpendicular to $2y + x = 8$ and that passes through the point B.

# Miscellaneous exercise four

This miscellaneous exercise may include questions involving the work of this chapter, the work of any previous chapters, and the ideas mentioned in the Preliminary work section at the beginning of the book.

**1** Which of the following are equations of straight lines?

A: $y = 2x + 3$    B: $y = x^2 + 3$    C: $y = 3x - 1$    D: $y = 3 - x^3$

E: $y = 4 - 3x$    F: $2y = 4x + 5$    G: $y = 2x + 3x^2$    H: $y = 4x$

I: $x = 4y$    J: $2y + 3x = 5$    K: $y = \dfrac{2}{x+1}$    L: $y = \dfrac{x+1}{2}$

**2** State whether each of the points $A$ to $E$ lie on the line $y = 3x - 5$.

A(6, 12),    B(5, 11),    C(2, 1),    D(–3, –13),    E(–1, –8)

**3** State whether each of the points $F$ to $J$ lie on the line $y = -x + 6$.

F(1, 5),    G(0, 6),    H(2, 8),    I(–1, 4),    J(6, 0)

**4** For $f(x) = 2x + 3$ and $g(x) = 5x - 18$ determine each of the following.

**a** $f(4)$    **b** $f(-2)$    **c** $f(10)$

**d** $g(2)$    **e** $g(-2)$    **f** $g(6.5)$

**g** $f(1) + f(2)$    **h** $g(1) + g(2)$    **i** $f(m) + g(m)$

**j** The value of $m$ given that $f(m) = 15$.    **k** The value of $p$ given that $g(p) = 7$.

**l** The value of $q$ given that $f(q) = g(q)$.    **m** The value of $r$ given that $f(r) = r$.

**n** The value of $s$ given that $g(s) = s$.

**5** Find where each of the following pairs of lines intersect

**a** $y = 2x - 11$ and $y = -3x + 4$    **b** $5x + 2y = 3$ and $2x + 3y = 10$

**6** State the natural domain and the corresponding range for each of the following.

**a** $f(x) = x - 5$    **b** $f(x) = \sqrt{x-5}$    **c** $f(x) = (x-5)^2$

**d** $f(x) = \dfrac{1}{x-5}$    **e** $f(x) = \dfrac{1}{(x-5)^2}$    **f** $f(x) = \dfrac{1}{\sqrt{x-5}}$

**7** Prove that points A(29, 16), B(25, 24), C(–2, 33), D(–10, 29), E(–15, –6), F(29, 2)

**a** all lie on the same circle centre O, coordinates (5, 9),

**b** are such that OA is perpendicular to OC,

**c** are such that BE is a diagonal of the circle.

**8** Given the relationship between $x$ and $y$ is linear, determine the values of $a, b, c, \dots g$.

| $x$ | 0 | 1 | 2 | 3 | 4 | 5 | 6 | | $f$ | $g$ |
|---|---|---|---|---|---|---|---|---|---|---|
| $y$ | $a$ | $b$ | $c$ | 14 | $d$ | 24 | $e$ | | 54 | 494 |

**9** Each vertex of an equilateral triangle of side 10 cm is the centre of a circle of radius 5 cm. Find the area of the central region bounded by the circles giving your answer as an exact value.

# 5.

# Quadratic functions

The *Preliminary work* reminded you of what the graph of $y = x^2$ looks like but what of the graphs of $y = x^2 + 3$ or $y = (x - 3)^2$ or $y = (x - 3)^2 + 4$, etc. Are you familiar with the effects changing $a$, $p$ and $q$ have on the graph of $y = a(x - p)^2 + q$?

If you think you can remember something about these things then continue on to the 'situation' described below. If you are not familiar with any of this work on the graphs of quadratic functions then read the next page and work through **Exercise 5A** first and then come back to the situation described below.

## Situation

An engineer needs to write instructions for a computer operated cutting machine.

The machine cuts four 'teeth' along one of the long sides of a rectangular piece of metal, 16 cm × 10 cm.

The instructions need to give the height of the teeth and the equations of the cuts required.

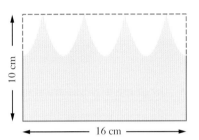

The particular job shown requires the teeth to be cut as shown on the right.

The engineer starts writing the instructions as follows:

'Height' of teeth 4 cm. For $0 \le x < 2$, cut $y = x^2$.

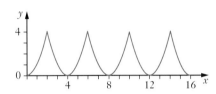

For the next instruction, i.e. for $2 \le x < 6$, the engineer is not sure which of the following equations to use:

$$y = x^2 + 4, \qquad y = x^2 - 4, \qquad y = (x - 4)^2, \qquad y = (x + 4)^2.$$

Decide which it should be and then write out the complete instructions for $0 \le x \le 16$.

The machine is now required to cut teeth as shown on the right, in a sheet of metal that is 18 cm long.

The engineer began her instructions:

'Height' of teeth 4 cm. For $0 \le x < 2$, cut $y = x^2$.

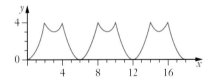

Again the engineer is not sure which equation to use for the next instruction, i.e. for

$$2 \le x < 4.$$

She knows that it should be one of the following:

$$y = (x - 3)^2 + 3, \qquad y = (x - 3)^2 - 3, \qquad y = (x + 3)^2 + 3, \qquad y = (x + 3)^2 - 3.$$

Decide which it should be and then write out the complete instructions for $0 \le x \le 18$.

# Quadratic functions

The points indicated in the table below would clearly *not* lie in a straight line because as the $x$ value increases by 1 the $y$ value does *not* alter by a constant amount.

| $x$ | $-3$ | $-2$ | $-1$ | 0 | 1 | 2 | 3 |
|---|---|---|---|---|---|---|---|
| $y$ | 9 | 4 | 1 | 0 | 1 | 4 | 9 |

As you probably recognised, the table is for $y = x^2$, the most basic quadratic function.

The graph of $y = x^2$ is shown on the right and is said to be **parabolic** in shape. The curve is a known as a **parabola**.

Note that:

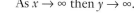

- the graph is symmetrical with the $y$-axis as the line of symmetry.

- the graph is shaped like 'a valley' coming down to a 'low point' called the **minimum** point.

- for larger and larger positive $x$ values, $y$ takes even larger positive values. We say that 'as $x$ approaches infinity then $y$ approaches infinity' (and indeed it does it faster than $x$ does). This is written:  As $x \to \infty$ then $y \to \infty$.

- Similarly:  As $x \to -\infty$ then $y \to \infty$.

All functions that can be written in the form

$$y = ax^2 + bx + c \quad \text{for } a \neq 0$$

will have graphs that are the same basic shape as that of $y = x^2$ but may be moved left, right, up, down, flipped over, squeezed or stretched.

Just as the equations of linear functions are not always presented in the form

$$y = mx + c$$

(for example $2y = 3x + 4$, $3x + 4y = 7$, $6x + y - 5 = 0$), then so the equations of quadratic functions are not always presented in the form

$$y = ax^2 + bx + c.$$

Indeed for quadratic functions there are three ways in which the rule is frequently given:

$$y = ax^2 + bx + c, \qquad \text{for example} \qquad y = 2x^2 - 12x + 10,$$
$$y = a(x - b)(x - c), \qquad \text{for example} \qquad y = 2(x - 1)(x - 5),$$
and $\qquad y = a(x - b)^2 + c, \qquad \text{for example} \qquad y = 2(x - 3)^2 - 8.$

The reader should confirm that by expanding the brackets and collecting like terms the last two forms can be written in the form $y = ax^2 + bx + c$, and indeed are just alternative ways of writing the first equation, $y = 2x^2 - 12x + 10$.

Whilst the letters $a$, $b$ and $c$ are quite commonly used this does not have to be the case. Exercise 5A which follows, considers quadratic functions with equations given in the form

$$y = a(x - p)^2 + q$$

and investigates how changing the values of $a$, $p$ and $q$ alters the graph of the function.

ISBN 9780170390330

## Exercise 5A

This exercise should refresh your memory as to the effect altering the values $a$, $p$ and $q$ have on the graph of $y = a(x - p)^2 + q$.

Either work through the exercise as given below or, alternatively, use the ability of some calculators or internet sites to automatically move through displays of $y = ax^2$, $y = x^2 + q$ or $y = (x - p)^2$ for changing values of $a$, $p$ and $q$ to explore these aspects.

**1 Changing the value of '$a$' in $y = ax^2$.**

Display the graphs of the following functions altogether on the screen of a graphic calculator using an $x$-axis from –4 to 4 and a $y$-axis from –10 to 10.

$$y = x^2, \qquad y = 2x^2, \qquad y = 4x^2, \qquad y = 0.5x^2, \qquad y = -2x^2.$$

Write a few sentences explaining the effect changing the value of $a$ has on the graph of $y = ax^2$, referencing your comments to how the graphs differ from that of $y = x^2$.

**2 Changing the value of '$q$' in $y = x^2 + q$.**

Display the graphs of the following functions altogether on the screen of a graphic calculator using an $x$-axis from –3 to 3 and a $y$-axis from –4 to 14.

$$y = x^2, \qquad y = x^2 + 2, \qquad y = x^2 + 4, \qquad y = x^2 - 3.$$

Write a few sentences explaining the effect changing the value of q has on the graph of $y = x^2 + q$. Reference your comments to how the various graphs differ from that of $y = x^2$.

**3 Changing the value of '$p$' in $y = (x - p)^2$.**

Display the graphs of the following functions altogether on the screen of a graphic calculator using an $x$-axis from –6 to 8 and a $y$-axis from –10 to 10.

$$y = x^2, \qquad y = (x - 2)^2, \qquad y = (x - 5)^2, \qquad y = (x + 3)^2.$$

Write a few sentences explaining the effect changing the value of $p$ has on the graph of $y = (x - p)^2$. Reference your comments to how the various graphs differ from that of $y = x^2$.

**4** Display the graphs of the two functions below on the screen of a graphic calculator using an $x$-axis from –6 to 6 and a $y$-axis from –8 to 16.

$$y = (x - 3)^2 + 4, \qquad y = 2(x + 1)^2 - 4$$

Check that the location of the graphs on the calculator display agrees with your expectations as a result of doing questions 1, 2 and 3.

iStock.com/urfinguss

# Key features of the graph of $y = a(x - p)^2 + q$

You should have discovered that the graph of $y = a(x - p)^2 + q$ has the same shape as that of $y = ax^2$ but has moved $p$ units to the right and $q$ units up.

Thus:

- whilst the graph of $y = ax^2$ has the $y$-axis as its line of symmetry, the graph of $y = a(x - p)^2 + q$ has the line $x = p$ as its line of symmetry.

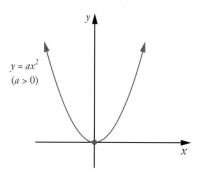
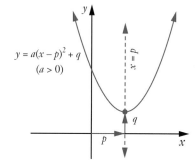

- whilst the graph of $y = ax^2$ has a turning point at $(0, 0)$, the graph of $y = a(x - p)^2 + q$ has a turning point at $(p, q)$.

  For $a > 0$ this will be a **minimum** turning point:

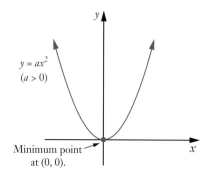
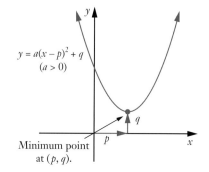

  Note: This 'valley shape' is sometimes referred to as being **concave up**.

  For $a < 0$ this will be a **maximum** turning point:

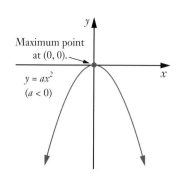
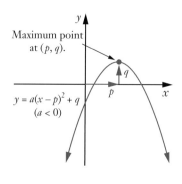

  Note: This 'hill shape' is sometimes referred to as being **concave down**.

ISBN 9780170390330

# Determining the rule from key features of the graph

## EXAMPLE 1

In the graph shown, the curves A, B and C are identical in shape to the curve shown in red, though B and C are 'upside down' versions. The red curve has equation $y = x^2$. Write down the equations of curves A, B and C.

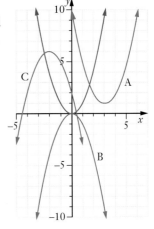

### Solution

Curve A is the red curve, $y = x^2$, moved right 3 and up 1.

Curve A has equation $y = (x - 3)^2 + 1$.

Curve B is the red curve, $y = x^2$, reflected in the $x$ axis.

Curve B has equation $y = -x^2$.

Curve C is curve B, $y = -x^2$, moved 2 units left and 6 units up.

Thus curve C has equation $y = -1(x + 2)^2 + 6$.

$\qquad\qquad$ I.e. $y = -(x + 2)^2 + 6$.

## EXAMPLE 2

Given that each of the following graphs show quadratic functions, determine the rule for each function.

a

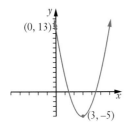

b

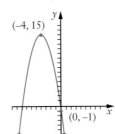

### Solution

a  Given that the graph shows a quadratic function with a turning point at $(3, -5)$ the equation
will be of the form $\qquad\qquad y = a(x - 3)^2 - 5$.

$(0, 13)$ must 'fit' the equation, hence $\quad 13 = a(0 - 3)^2 - 5$,

so $\qquad\qquad\qquad\qquad\qquad 13 = 9a - 5$.

Solving gives $\qquad\qquad\qquad a = 2$.

The given graph has equation $y = 2(x - 3)^2 - 5$.

b  Given that the graph shows a quadratic function with a turning point at $(-4, 15)$ the equation
will be of the form $\qquad\qquad y = a(x + 4)^2 + 15$.

(And because of the *maximum* turning point we expect $a$ to be negative.)

$(0, -1)$ must 'fit' the equation, hence $\quad -1 = a(0 + 4)^2 + 15$,

so $\qquad\qquad\qquad\qquad\qquad -16 = 16a$.

Hence $\qquad\qquad\qquad\qquad a = -1$.

The given graph has equation $y = -(x + 4)^2 + 15$.

## Exercise 5B

**1** In the graphs below, the curves A, B, C, D, E and F are identical in shape to the red curve shown, which has equation $y = x^2$. Write down the equations of curves A, B, C, D, E and F.

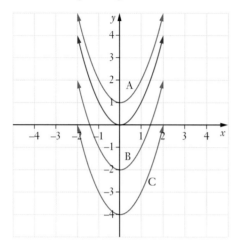 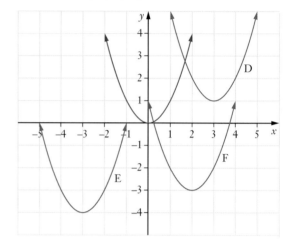

**2** The curves G, H, I and J are all identical in shape to the red curve shown (but are each 'upside down' versions). If the red curve has equation $y = x^2$, write down the equations of curves G, H, I and J.

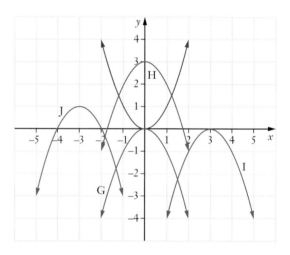

**3** The curves K, L, M and N are all of the same shape as the red curve shown. If the red curve has equation $y = 2x^2$, write down the equations of curves K, L, M and N.

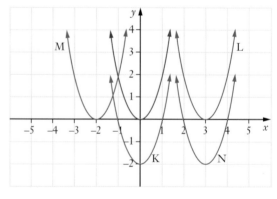

ISBN 9780170390330

**4** Given that each of the following graphs show quadratic functions, determine the equation of each one.

**a**

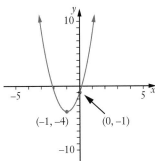

**b**

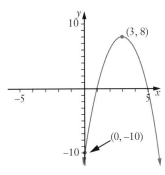

**c**

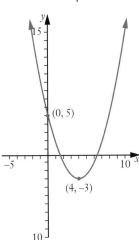

**d**

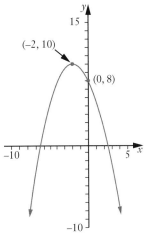

# Determining the key features of the graph from the rule

Each of the three ways the rules for quadratic functions are usually given, i.e.:

$$y = ax^2 + bx + c, \qquad y = a(x - b)(x - c), \qquad y = a(x - b)^2 + c$$

readily provide information about some of the key features of the graph of the function.

Note: In the expression $ax^2 + bx + c$, $a$ is said to be the **coefficient** of $x^2$, and $b$ is said to be the **coefficient** of $x$.

## Rule given in the form $y = a(x - b)^2 + c$

- *Line of symmetry*

   The graph of $y = a(x - b)^2 + c$ has the line $x = b$ as its line of symmetry.

- *Turning point*

   For $a > 0$, $y = a(x - b)^2 + c$ has a minimum turning point at $(b, c)$.

   For $a < 0$, $y = a(x - b)^2 + c$ has a maximum turning point at $(b, c)$.

- *y-axis intercept*

   All points on the $y$-axis have an $x$-coordinate of zero. Hence substituting $x = 0$ into $y = a(x - b)^2 + c$ determines the $y$-axis intercept.

EXAMPLE 3

For the graph of $y = 2(x - 2)^2 - 5$, determine

a the equation of the line of symmetry

b the coordinates of the maximum/minimum turning point, stating which of these it is

c the coordinates of the point where it cuts the $y$-axis

d Show these features on a sketch of the graph of $y = 2(x - 2)^2 - 5$.

**Solution**

a The quadratic function $y = a(x - b)^2 + c$ has line of symmetry $x = b$.
Hence $y = 2(x - 2)^2 - 5$ has line of symmetry $x = 2$.

b For $a > 0$, $y = a(x - b)^2 + c$ has a minimum turning point at $(b, c)$.
Hence $y = 2(x - 2)^2 - 5$ has a minimum turning point at $(2, -5)$.

c All points on the $y$-axis have an $x$-coordinate of zero.

Given $\qquad y = 2(x - 2)^2 - 5$,

when $x = 0$, $\qquad y = 2(0 - 2)^2 - 5$

$\qquad\qquad\qquad = 3.$

The graph of $y = 2(x - 2)^2 - 5$ cuts the $y$-axis at $(0, 3)$.

d Placing the information from the previous parts on a graph, below left, the sketch can be completed, below right. (Note how the point symmetrical with $(0, 3)$ has been included to help with the sketch.)

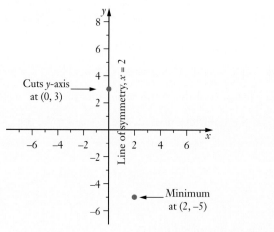
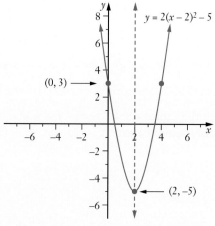

Note: When a question requires a *sketch* of a graph to be made, it does not mean neatness and reasonable accuracy can be ignored. We would not expect to be able to rely on great accuracy of values read from a sketch graph but the sketch should be neat, reasonably accurate and with the noteworthy features of the graph evident.

ISBN 9780170390330

- The reader can confirm the correctness of the previous sketch by viewing the graph of

$$y = 2(x - 2)^2 - 5$$

on a graphic calculator.

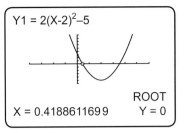

Y1 = 2(X-2)²–5

ROOT
X = 0.418 861 169 9    Y = 0

- The calculator display shown also indicates that the graph cuts the $x$-axis at $(0.418\,861\,169\,9, 0)$. Asked to determine the other $x$-axis intercept the calculator gives $(3.581\,138\,83, 0)$.

These values can also be obtained algebraically as follows.

All points on the $x$-axis have a $y$-coordinate of zero. Thus for $y = 2(x - 2)^2 - 5$,

| | |
|---|---|
| when $y = 0$, | $0 = 2(x - 2)^2 - 5$ |
| Add 5 to each side: | $5 = 2(x - 2)^2$ |
| Divide each side by 2: | $2.5 = (x - 2)^2$ |
| Hence | $x - 2 = \pm\sqrt{2.5}$ |

$$x = 2 + \sqrt{2.5} \quad \text{or} \quad 2 - \sqrt{2.5}$$
$$= 3.58 \quad \text{or} \quad 0.42 \ (2 \text{ decimal places}).$$

Notice that whilst these $x$-axis intercepts were not immediately obvious from the equation, a very acceptable sketch was possible without knowing them. Indeed approximate values for these intercepts could have been obtained from the sketch on the previous page.

## EXAMPLE 4

For the graph of $y = -(x - 3)^2 + 5$ determine

a    the equation of the line of symmetry

b    the coordinates of the maximum/minimum turning point, stating which of these it is

c    the coordinates of the point where $y = -(x - 3)^2 + 5$ cuts the $y$-axis.

d    Show these features on a sketch of the graph of $y = -(x - 3)^2 + 5$.

### Solution

a    The quadratic function $y = a(x - b)^2 + c$ has line of symmetry $x = b$.

Hence $y = -(x - 3)^2 + 5$ has line of symmetry $x = 3$.

b    For $a < 0$, $y = a(x - b)^2 + c$ has a maximum turning point at $(b, c)$.

$y = -(x - 3)^2 + 5$ has $a = -1$ and so has a maximum turning point at $(3, 5)$.

c    All points on the $y$-axis have an $x$-coordinate of zero.

| | |
|---|---|
| Given | $y = -(x - 3)^2 + 5$, |
| when $x = 0$, | $y = -(0 - 3)^2 + 5$ |
| | $= -4$. |

The graph of $y = -(x - 3)^2 + 5$ cuts the $y$-axis at $(0, -4)$.

**d** Placing the information from the previous parts on a graph, below left, the sketch can be completed, below right. (Note the inclusion of the point symmetrical with (0, –4) to help with the sketch.)

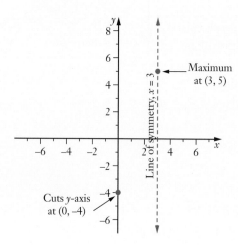

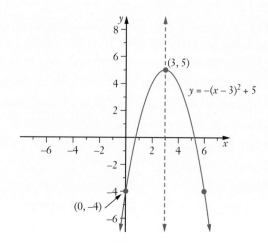

## Rule given in the form $y = a(x - b)(x - c)$

- *y-axis intercept*

  All points on the $y$-axis have an $x$-coordinate of zero. Hence substituting $x = 0$ into the equation $y = a(x - b)(x - c)$ allows the $y$-axis intercept to be determined.

- *x-axis intercepts*

  If $x = b$,  $y = a(b - b)(b - c)$        If $x = c$,  $y = a(c - b)(c - c)$
  
  $\qquad\qquad = a(0)(b - c)$        $\qquad\qquad\qquad = a(c - b)(0)$
  
  $\qquad\qquad = 0$        $\qquad\qquad\qquad\qquad = 0$

  But if a point has a $y$-coordinate of zero it must lie on the $x$-axis.

  Hence the graph of $y = a(x - b)(x - c)$ cuts the $x$-axis at $(b, 0)$ and at $(c, 0)$.

- *Line of symmetry*

  With the curve cutting the $x$-axis at $(b, 0)$ and $(c, 0)$ the line of symmetry must cut the $x$-axis at a point mid-way between these two points. This allows the equation of the line of symmetry to be determined.

- *Turning point*

  The turning point must lie on the line of symmetry and:

  for $a > 0$, $y = a(x - b)(x - c)$ will have a **minimum** turning point,

  for $a < 0$, $y = a(x - b)(x - c)$ will have a **maximum** turning point.

EXAMPLE 5

For the graph of $y = (x + 1)(x - 5)$ determine

a   the coordinates of the $y$-axis intercept
b   the coordinates of the $x$-axis intercepts
c   the equation of the line of symmetry
d   the nature and location of the turning point.
e   Sketch the curve $y = (x + 1)(x - 5)$.

**Solution**

a   For every point on the $y$-axis, $x = 0$.

  If $x = 0$,  $y = (0 + 1)(0 - 5)$
  $= (1)(-5)$
  $= -5$

  The curve cuts the $y$-axis at the point $(0, -5)$.

b   For every point on the $x$-axis, $y = 0$.

  This will occur when   $x + 1 = 0$   and when   $x - 5 = 0$,
  i.e. when              $x = -1$      and when   $x = 5$.

  The curve cuts the $x$-axis at the points $(-1, 0)$ and $(5, 0)$.

c   The line of symmetry will cut the $x$-axis at the point mid-way between $(-1, 0)$ and $(5, 0)$,
  i.e. the point $(2, 0)$.

  The line of symmetry is the line $x = 2$.

d   $y = a(x - b)(x - c)$ has a *minimum* turning point when $a > 0$.

  For $y = (x + 1)(x - 5)$, $a = 1$ so we have a minimum turning point.

  The turning point must lie on the line of symmetry.

  Hence its $x$-coordinate must equal 2.

  If $x = 2$,  $y = (2 + 1)(2 - 5)$
  $= (3)(-3)$
  $= -9$.

  The curve has a minimum turning point at $(2, -9)$.

e   Placing the information from the previous parts on a graph, below left, the sketch can be completed, below right. (Note how the point symmetrical with $(0, -5)$ has been included to help with the sketch.)

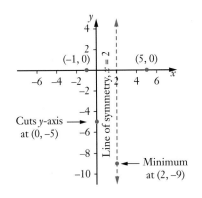
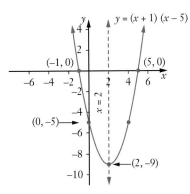

## Rule given in the form $y = ax^2 + bx + c$

- *Line of symmetry*

  The graph of $y = ax^2 + bx + c$ has the line $x = -\dfrac{b}{2a}$ as its line of symmetry.

  This claim can be justified as follows:

  The graph of $y = ax^2 + bx + c$ is that of $\quad y = ax^2 + bx$ moved up $c$ units.

  Now $y = ax^2 + bx$ cuts the $x$-axis when $\quad 0 = ax^2 + bx$,

  $$\text{i.e.} \quad 0 = x(ax + b).$$

  Hence the $x$-axis intercepts will be $\quad x = 0 \quad$ and $\quad x = -\dfrac{b}{a}$.

  The line of symmetry of $y = ax^2 + bx$, and hence of $y = ax^2 + bx + c$, will pass through the midpoint of

  these $x$-axis intercepts, which is $-\dfrac{b}{2a}$.

- *Turning point*

  The turning point must lie on the line of symmetry and:

  for $a > 0$, $\quad y = ax^2 + bx + c \quad$ this will be a **minimum** turning point,

  and $\quad$ for $a < 0$, $\quad y = ax^2 + bx + c \quad$ this will be a **maximum** turning point.

- *y-axis intercept*

  All points on the $y$-axis have an $x$-coordinate of zero. Substituting $x = 0$ into the equation $y = ax^2 + bx + c$ allows the $y$-axis intercept to be determined as $(0, c)$.

## EXAMPLE 6

For the graph of $y = 2x^2 - 6x + 1$ determine

a   the line of symmetry

b   the nature and location of the turning point

c   the $y$-axis intercept coordinates.

d   Hence sketch the curve.

### Solution

a   The graph of $y = ax^2 + bx + c$ has the line $x = -\dfrac{b}{2a}$ as its line of symmetry.

$y = 2x^2 - 6x + 1$ has $a = 2$, $b = -6$ and $c = 1$.

The line of symmetry is $x = 1.5$.

b   With the coefficient of $x^2$ being positive, we have a minimum turning point.

The turning point must lie on the line of symmetry. Thus $x = 1.5$.

With $x = 1.5$, $\quad y = 2(1.5)^2 - 6(1.5) + 1$

$$= -3.5.$$

The turning point is a minimum and occurs at $(1.5, -3.5)$.

c   Every point on the $y$-axis has an $x$-coordinate of zero.

With $x = 0$, $\quad y = 2(0)^2 - 6(0) + 1$

$$= 1$$

The $y$-axis intercept has coordinates $(0, 1)$.

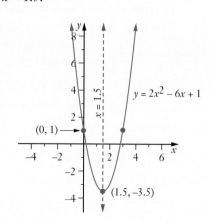

d   The sketch can be completed, as shown on the right.
(Note how the point symmetrical with $(0, 1)$ has been included to help with the sketch.)

## Exercise 5C

Rule given in the form $y = a(x - p)^2 + q$

**1** For the graph of $y = (x + 1)^2 - 4$ determine

   **a**   the equation of the line of symmetry

   **b**   the coordinates of the maximum/minimum turning point, stating which of these it is

   **c**   the coordinates of the point where $y = (x + 1)^2 - 4$ cuts the $y$-axis.

   **d**   Show these features on a sketch of $y = (x + 1)^2 - 4$.

**2** For the graph of $y = (x - 3)^2 + 5$ determine

   **a**   the equation of the line of symmetry

   **b**   the coordinates of the maximum/minimum turning point, stating which of these it is

   **c**   the coordinates of the point where $y = (x - 3)^2 + 5$ cuts the $y$-axis.

   **d**   Show these features on a sketch of $y = (x - 3)^2 + 5$.

**3** For the graph of $y = -2(x - 1)^2 + 3$ determine

   **a**   the equation of the line of symmetry

   **b**   the coordinates of the maximum/minimum turning point, stating which of these it is

   **c**   the coordinates of the point where $y = -2(x - 1)^2 + 3$ cuts the $y$-axis.

   **d**   Show these features on a sketch of $y = -2(x - 1)^2 + 3$.

Rule given in the form $y = a(x - b)(x - c)$

**4** For the graph of $y = (x - 3)(x - 7)$ determine

   **a**   the coordinates of the point where the curve cuts the $y$-axis

   **b**   the coordinates of the points where the curve cuts the $x$-axis

   **c**   the equation of the line of symmetry

   **d**   the nature and location of the turning point.

   **e**   Show these features on a sketch of $y = (x - 3)(x - 7)$.

**5** For the graph of $y = (x - 3)(x + 4)$ determine

   **a**   the coordinates of the point where the curve cuts the $y$-axis

   **b**   the coordinates of the points where the curve cuts the $x$-axis

   **c**   the equation of the line of symmetry

   **d**   the nature and location of the turning point.

   **e**   Show these features on a sketch of $y = (x - 3)(x + 4)$.

**6** For the graph of $y = (x + 2)(x + 4)$ determine

   **a**   the coordinates of the point where the curve cuts the $y$-axis

   **b**   the coordinates of the points where the curve cuts the $x$-axis

   **c**   the equation of the line of symmetry

   **d**   the nature and location of the turning point.

   **e**   Show these features on a sketch of $y = (x + 2)(x + 4)$.

**7** For the graph of $y = x^2 + 4x - 12$ determine

  **a** the equation of the line of symmetry

  **b** the location and nature of the turning point

  **c** the coordinates of the $y$-axis intercept.

  **d** Hence sketch the curve.

**8** For the graph of $y = x^2 - 6x + 1$ determine

  **a** the equation of the line of symmetry

  **b** the location and nature of the turning point

  **c** the coordinates of the $y$-axis intercept.

  **d** Hence sketch the curve.

iStock.com/Gannet77

**9** For the graph of $y = -2x^2 + 4x + 1$ determine

  **a** the equation of the line of symmetry

  **b** the location and nature of the turning point

  **c** the coordinates of the $y$-axis intercept.

  **d** Hence sketch the curve.

**10** For the graph of $y = 8x - 2x^2 - 3$ determine

  **a** the equation of the line of symmetry

  **b** the location and nature of the turning point

  **c** the coordinates of the $y$-axis intercept.

  **d** Hence sketch the curve.

## Applications

**11** A dog owner wishes to enclose a rectangular area in his backyard for the dog. The owner wishes to use existing fencing along two adjacent sides of the rectangle and has 14 metres of new fencing available for the other two sides. Suppose that we let the dimensions of the rectangle be $x$ metres and $y$ metres and the area be $A$ m$^2$ as shown.

Existing Fence

$x$ metres

Area, $A$ m$^2$

Existing Fence

$y$ metres

  **a** Explain why $\quad x + y = 14$.

  **b** Explain why $\quad A = x(14 - x)$.

  **c** With $A$ on the vertical axis and $x$ on the horizontal axis make a sketch of
$$A = x(14 - x).$$

  **d** What is the greatest rectangular area the owner can enclose and what would be the dimensions that give this greatest area?

ISBN 9780170390330

**12** A gardener has 20 metres of fencing and wishes to fence off a rectangular area using the twenty metres of fencing on three sides and an existing fence on the fourth. Suppose we define the variables $x$, $y$ and $A$ as shown in the diagram.

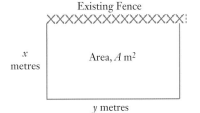

**a** Explain why $A = 20x - 2x^2$.

**b** With $A$ on the vertical axis and $x$ on the horizontal axis make a sketch of
$$A = 20x - 2x^2.$$

**c** What is the greatest rectangular area the gardener can enclose and what would be the dimensions that give this greatest area?

**For the following questions view the graph on your calculator if you wish.**

**13** The diagram below shows the motion of a particle fired with initial speed of 7 m/s at 45° to the horizontal, from a position that is $h$ metres above ground level.

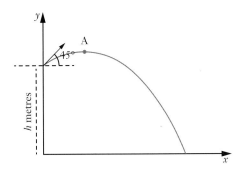

The equation of the path of the particle is
$$y = -0.2(x - 2.5)^2 + 11.25$$
with $x$ and $y$ axes as shown in the diagram, and 1 metre to 1 unit on each axis. Find

**a** the coordinates of $A$, the highest point on the path of the particle

**b** the value of $h$.

**c** State whether the graph shown is concave up or concave down.

**14** A housing market analyst was trying to predict the best time to purchase a house during a slump in sales and the consequent fall in prices. She felt that the decline in prices and the rise that was expected to follow would be according to the rule:
$$\text{Average house price (in \$1000s)} = 0.6t^2 - 12t + 590$$
where $t$ is the time in months.

If her quadratic model of the situation is correct

**a** what is the average house price when $t = 0$?

**b** what is the average house price when $t = 15$?

**c** when (i.e for what value of $t$) should she purchase a house and what should she expect the average price to be then?

**15** The height above ground, $h$ metres, of a stone projected vertically upwards, from ground level with initial speed 49 m/s, is given by $h = 49t - 4.9t^2$, where $t$ is the time in seconds after projection. Determine the maximum value of $h$ and the value of $t$ for which it occurs.

**16**

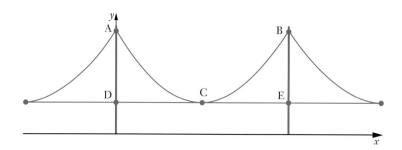

The diagram above shows a symmetrical suspension bridge over a river. The $x$-axis is the water level, the $y$-axis is as indicated, 1 metre is 1 unit on each axis and the support wire from A to B has the equation $y = \dfrac{3}{160}(x - 40)^2 + 15$.

**a**   Is the curve from A to B concave up or concave down?

**b**   How high is the bridge above the water level (i.e. how high is point C above the $x$-axis)?

**c**   Find the equation of the line of symmetry of the quadratic curve ACB.

How far is it from

**d**   D to C?   **e**   D to E?   **f**   D to A?

**17**   A road bridge is to be constructed over a tidal river. The road arch is in the shape of a quadratic function, as is the supporting arch (see diagram).

With $x$- and $y$- axes as shown the equations of each arch are as follows.

Road arch:   $y = -\dfrac{x^2}{2250} + \dfrac{2x}{15} + 40$

Supporting arch:   $y = \dfrac{2x}{3} - \dfrac{x^2}{450}$

Road Arch

Supporting Arch

(1 metre to 1 unit on each axis.)

The $x$-axis is the mean water level with the high and low tide levels being four metres either side of this mean.

**a**   Are the road arch and supporting arch concave up or concave down?

**b**   Calculate the value of $x$ at the mid-point of the bridge.

**c**   Calculate the length of the vertical strut between the support arch and the road arch at a point one quarter of the way along the bridge.

**d**   Calculate the maximum clearance between the water and the bridge at

  **i**   low tide   **ii**   high tide.

# Tables of values

As the Preliminary section mentioned, quadratic functions have tables of values with a *constant second difference pattern*.

For example, for $y = 2x^2 - 6x + 1$:

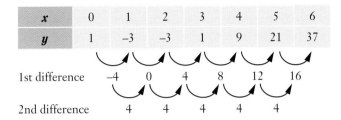

This constant second difference becomes clear when we consider a table of values for the general quadratic $y = ax^2 + bx + c$:

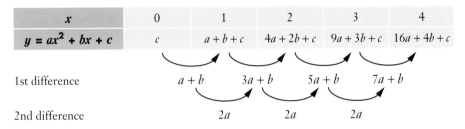

Consider the following table of values with, 1st and 2nd difference patterns as shown:

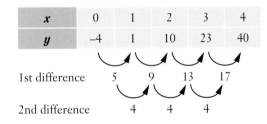

- The constant second difference pattern confirms that the relationship is quadratic.

- Comparing the constant second difference of 4 with $2a$, the constant second difference pattern in the table for the general quadratic, $y = ax^2 + bx + c$, gives us that $a = 2$.

- From the table for the general quadratic we see that when $x = 0$, $y = c$. Thus $c = -4$.

- Comparing the $y$ values for $x = 1$:    $a + b + c = 1$

    Thus, with $a = 2$ and $c = -4$,    $b = 3$.

The table of values is for the quadratic $y = 2x^2 + 3x - 4$.

ISBN 9780170390330

Consider the table and difference patterns shown below:

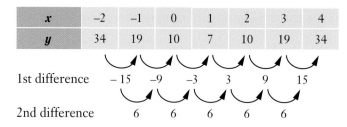

| $x$ | −2 | −1 | 0 | 1 | 2 | 3 | 4 |
|---|---|---|---|---|---|---|---|
| $y$ | 34 | 19 | 10 | 7 | 10 | 19 | 34 |

1st difference   − 15   −9   −3   3   9   15

2nd difference       6   6   6   6   6

The constant second difference pattern confirms that the table is for a quadratic function.

To determine the equation of the function we could compare the given table with that of the table for $y = ax^2 + bx + c$, as we did on the previous page, to determine that

$$2a = 6, \qquad\qquad c = 10, \qquad\qquad a + b + c = 7.$$

Hence $\qquad\qquad a = 3, \qquad\qquad c = 10, \qquad\qquad b = -6.$

The rule is $y = 3x^2 - 6x + 10$.

Alternatively, from the symmetrical nature of the $y$ values, the line of symmetry is $x = 1$ and the minimum point has coordinates $(1, 7)$. Thus the rule is of the form

$$y = a(x - 1)^2 + 7.$$

The point $(0, 10)$ must 'fit'. Thus $\qquad 10 = a(0 - 1)^2 + 7,$

giving $\qquad\qquad a = 3.$

The rule is $\qquad\qquad y = 3(x - 1)^2 + 7.$

The reader should confirm that these answers, $\quad y = 3x^2 - 6x + 10$

and $\qquad\qquad\qquad y = 3(x - 1)^2 + 7$, are indeed the same.

## Exercise 5D

For each of the tables shown in questions 1 to 12, by considering difference patterns, determine whether the relationship between $x$ and $y$ is linear, quadratic or neither of these. For those relationships that are either linear or quadratic determine the rule.

**1**
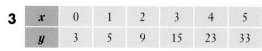

| $x$ | 0 | 1 | 2 | 3 | 4 | 5 |
|---|---|---|---|---|---|---|
| $y$ | 5 | 12 | 21 | 32 | 45 | 60 |

**2**
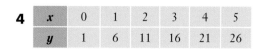

| $x$ | 0 | 1 | 2 | 3 | 4 | 5 |
|---|---|---|---|---|---|---|
| $y$ | 0 | 1 | 8 | 27 | 64 | 125 |

**3**

| $x$ | 0 | 1 | 2 | 3 | 4 | 5 |
|---|---|---|---|---|---|---|
| $y$ | 3 | 5 | 9 | 15 | 23 | 33 |

**4**

| $x$ | 0 | 1 | 2 | 3 | 4 | 5 |
|---|---|---|---|---|---|---|
| $y$ | 1 | 6 | 11 | 16 | 21 | 26 |

**5**

| $x$ | 0 | 1 | 2 | 3 | 4 | 5 |
|---|---|---|---|---|---|---|
| $y$ | 2 | 3 | 6 | 11 | 18 | 27 |

**6**

| $x$ | 0 | 1 | 2 | 3 | 4 | 5 |
|---|---|---|---|---|---|---|
| $y$ | $\pi$ | $2\pi$ | $3\pi$ | $4\pi$ | $5\pi$ | $6\pi$ |

ISBN 9780170390330

**7**

| $x$ | 0 | 1 | 2 | 3 | 4 | 5 |
|---|---|---|---|---|---|---|
| $y$ | 3 | 6 | 12 | 24 | 48 | 96 |

**8**

| $x$ | 4 | 3 | 0 | 5 | 1 | 2 |
|---|---|---|---|---|---|---|
| $y$ | 40 | 28 | 4 | 54 | 10 | 18 |

**9**

| $x$ | 1 | 4 | 2 | 0 | 5 | 3 |
|---|---|---|---|---|---|---|
| $y$ | 11 | 35 | 19 | 3 | 43 | 27 |

**10**

| $x$ | 1 | 3 | 2 | 5 | 4 | 0 |
|---|---|---|---|---|---|---|
| $y$ | 5 | 21 | 11 | 53 | 35 | 3 |

**11**

| $x$ | –1 | 0 | 1 | 2 | 3 | 4 | 5 |
|---|---|---|---|---|---|---|---|
| $y$ | 28 | 13 | 4 | 1 | 4 | 13 | 28 |

**12**

| $x$ | –2 | 0 | 2 | 4 | 6 | 8 |
|---|---|---|---|---|---|---|
| $y$ | –20 | –4 | 4 | 4 | –4 | –20 |

**13**

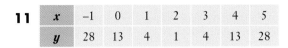

**a** Copy and complete the table shown below for the pattern shown above.

| Length of side of cube ($L$ units) | 1 | 2 | 3 | 4 | 5 | 6 |
|---|---|---|---|---|---|---|
| Surface area of cube ($n$ units$^2$) | 6 | 24 | | | | |

**b** Use difference patterns to determine whether the relationship between $L$ and $n$ is linear, quadratic or neither of these.

**c** If linear or quadratic, determine the rule.

**14**

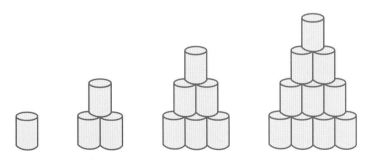

**a** Copy and complete the table shown below for the pattern shown above.

| Number of rows of cans ($r$) | 1 | 2 | 3 | 4 | 5 | 6 |
|---|---|---|---|---|---|---|
| Number of cans ($n$) | 1 | 3 | 6 | | | |

**b** Use difference patterns to determine whether the relationship between $r$ and $n$ is linear, quadratic or neither of these.

**c** If linear or quadratic, determine the rule.

ISBN 9780170390330

## Social networks

Two computers in a network can involve 1 connection.

Three computers in a network can involve 3 connections.

Four computers in a network can involve 6 connections.

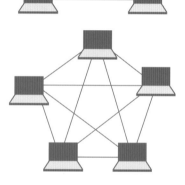

Five computers in a network can involve 10 connections.

How many connections can be involved in a six computer network?

How many connections can be involved in a seven computer network?

Investigate.

## Projectiles

What has the path of a projectile got to do with parabolas and quadratic functions?

ISBN 9780170390330

## Transmitters and receiving dishes

Why is the parabolic shape of significance for car headlights, receiving dishes and transmitters?

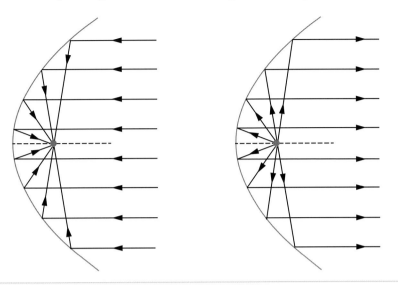

## Stopping distances

Suppose that a vehicle travelling at a particular speed requires $y$ metres to come to rest once normal braking is applied.

Under the same conditions, if the vehicle is travelling twice as fast as previously it will require not $2y$ but $4y$ metres to come to rest once normal braking is applied. I.e. the braking distance required and the distance travelled are *not* linearly related – double the speed, quadruple the distance required.

When something occurs ahead of a moving vehicle, requiring the driver to apply the brakes and come to a stop, the vehicle will initially travel on whilst the driver reacts to the situation, and then the vehicle will travel further 'under braking', until it is brought to rest.

Suppose that for a particular vehicle and road conditions the reaction distance, braking distance and total stopping distance at various speeds are as follows:

| Speed ($x$ km/h) | 40 | 50 | 60 | 70 | 80 | 90 | 100 | 110 |
|---|---|---|---|---|---|---|---|---|
| Reaction distance ($R$ m) | 12 | 15 | 18 | 21 | 24 | 27 | 30 | 33 |
| Braking distance ($B$ m) | 16 | 25 | 36 | 49 | 64 | 81 | 100 | 121 |
| Total stopping distance ($T$ m) | 28 | 40 | 54 | 70 | 88 | 108 | 130 | 154 |

What functions and rules are involved here?

# Completing the square

We saw earlier in this chapter that when a quadratic function is given in the form

$$y = a(x - p)^2 + q,$$

it is an easy matter to determine the line of symmetry, $x = p$, and the coordinates of the turning point, $(p, q)$.

For example the graph of the quadratic function

$$y = 2(x - 5)^2 - 7$$

has $x = 5$ as its line of symmetry and has a turning point (a minimum) at $(5, -7)$.

Given $\qquad\qquad\qquad\qquad y = 2(x - 5)^2 - 7$

and expanding: $\qquad\qquad\quad y = 2(x - 5)(x - 5) - 7$

$$= 2(x^2 - 10x + 25) - 7$$

$$= 2x^2 - 20x + 43$$

we obtain the rule in its $y = ax^2 + bx + c$ form.

$$y = 2x^2 - 20x + 43$$

However, how would we go about changing a quadratic rule given in the form

$$y = ax^2 + bx + c$$

to the form $\qquad\qquad\qquad\qquad y = a(x - p)^2 + q?$

The answer is that we use the technique referred to as 'completing the square'.

To use this method we need to be familiar with the expansion:

$$(x \pm e)^2 = x^2 \pm 2ex + e^2.$$

Note especially that the number on the end, $e^2$, is *the square of half the coefficient of* $x$.

---

## EXAMPLE 7

Express $y = x^2 + 3x - 5$ in the form $y = a(x - p)^2 + q$.

### Solution

Given $\qquad\qquad y = x^2 + 3x - 5,$

First create a gap to allow 'the square of half the coefficient of $x$' to be inserted.

$$y = x^2 + 3x \qquad\quad - 5$$

Then add, and then subtract, $\left(\dfrac{3}{2}\right)^2$, which is *the square of half the coefficient of* $x$.

$$y = x^2 + 3x + \left(\frac{3}{2}\right)^2 - 5 - \left(\frac{3}{2}\right)^2$$

$$\therefore \qquad\qquad y = \left(x + \frac{3}{2}\right)^2 - \frac{29}{4}$$

---

ISBN 9780170390330

EXAMPLE 8

Express

a $\quad y = x^2 + 6x - 3$            b $\quad y = 2x^2 - 4x + 1$

in the form $y = a(x - p)^2 + q$, and hence determine the nature and location of the turning point of the graph of the function.

**Solution**

a
$$\begin{aligned}
y &= x^2 + 6x - 3 \\
&= x^2 + 6x \quad\quad - 3 \\
&= x^2 + 6x + 9 - 3 - 9 \\
\therefore \quad y &= (x + 3)^2 - 12
\end{aligned}$$
Minimum turning point at $(-3, -12)$.

b
$$\begin{aligned}
y &= 2x^2 - 4x + 1 \\
&= 2(x^2 - 2x) \quad\quad + 1 \\
&= 2(x^2 - 2x + 1 - 1) + 1 \\
&= 2(x^2 - 2x + 1) - 2 + 1 \\
\therefore \quad y &= 2(x - 1)^2 - 1
\end{aligned}$$
Minimum turning point at $(1, -1)$.

## Exercise 5E

Express each of the following in the form $y = a(x - p)^2 + q$ and hence determine the nature and location of the turning point of the graph of each function.

**1** $\quad y = x^2 + 4x - 1$        **2** $\quad y = x^2 - 6x + 2$        **3** $\quad y = x^2 - 8x + 10$

**4** $\quad y = x^2 + 6x + 3$        **5** $\quad y = x^2 - 3x + 2$        **6** $\quad y = x^2 - 5x + 3$

**7** $\quad y = -x^2 + 10x - 1$        **8** $\quad y = 2x^2 - 12x + 3$        **9** $\quad y = -2x^2 + 8x + 4$

**10** $\quad y = 2x^2 + 5x + 4$

If we were to be repeatedly carrying out a procedure like completing the square we could carry out the process on the general quadratic $y = ax^2 + bx + c$ in order to obtain a formula. Then, given a specific quadratic, we simply substitute the appropriate values of $a$, $b$ and $c$.

$$\begin{aligned}
y &= ax^2 + bx + c \\
&= a\left(x^2 + \frac{b}{a}x\right) \quad\quad\quad\quad\quad + c \\
&= a\left(x^2 + \frac{b}{a}x + \frac{b^2}{4a^2} - \frac{b^2}{4a^2}\right) + c \\
&= a\left(x^2 + \frac{b}{a}x + \frac{b^2}{4a^2}\right) - \frac{b^2}{4a} \quad + c \\
&= a\left(x + \frac{b}{2a}\right)^2 + c - \frac{b^2}{4a}
\end{aligned}$$

Thus the expression $y = ax^2 + bx + c$ can be written as $y = a\left(x + \dfrac{b}{2a}\right)^2 + c - \dfrac{b^2}{4a}$

The reader should now repeat Example 8 above by substituting appropriate values for $a$, $b$ and $c$ into the formula above and check that the answers obtained are the same as those given by the working at the top of the page.

# Miscellaneous exercise five

This miscellaneous exercise may include questions involving the work of this chapter, the work of any previous chapters, and the ideas mentioned in the Preliminary work section at the beginning of the book.

**1** If $f(x) = 5x + 1$ and $g(x) = x^2 - 3$ determine

   **a** $f(6)$                     **b** $g(2)$                      **c** $f(2) + g(6)$.

**2** Classify each of the curves shown below as either 'concave up' or as 'concave down'.

   **a** 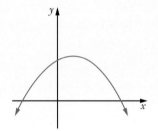       **b** 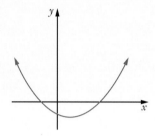       **c**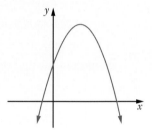

**3** Given that all of the points A to F given below lie on the line $y = 2x - 5$, determine the values of $a, b, c, d, e$ and $f$.

   A(3, $a$),        B(2, $b$),        C(–4, $c$),        D(2.5, $d$),        E($e$, 13),        F($f$, –5).

**4** **a** Find the gradient of a line that is perpendicular to a line that has a gradient of 2.

   **b** Find the gradient of a line that is perpendicular to $y = 3x - 4$.

   **c** Find the gradient of a line that is perpendicular to $y = -0.2x + 1$.

   **d** Find the equation of a line that passes through the point (3, 13) and is perpendicular to $y = -0.5x + 1$.

**5** For the graph of $y = (x - 1)(x - 3)$, determine

   **a** the coordinates of the $y$-axis intercept

   **b** the coordinates of the $x$-axis intercepts

   **c** the equation of the line of symmetry

   **d** the nature and location of the turning point.

**6** Copy and complete the following table.

| Equation | Cuts $y$-axis | Line of symmetry | Turning point | |
|---|---|---|---|---|
| | | | Coordinates | Max or min? |
| $y = x^2 + 4x + 1$ | (?, ?) | $x = ?$ | (?, ?) | |
| $y = x^2 - 2x - 1$ | (?, ?) | $x = ?$ | (?, ?) | |
| $y = 2x^2 + 4x - 3$ | (?, ?) | $x = ?$ | (?, ?) | |
| $y = 2x^2 + 6x - 1$ | (?, ?) | $x = ?$ | (?, ?) | |

**7** For the graph of $y = 2(x + 3)^2 - 4$ determine

**a** the equation of the line of symmetry

**b** the coordinates of the turning point.

The graph of $y = 2(x + 3)^2 - 4$ is moved 2 units to the right and 3 units up.

**c** What will be the equation of the line of symmetry of the 'new' curve?

**d** What will be the coordinates of the turning point of the 'new' curve?

**8** Determine the equation of each of the straight lines A to I shown below.

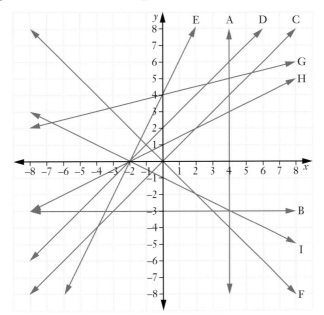

**9** The four curves I, II, III and IV shown on the right have their equations amongst the six listed below.

Select the appropriate equation for each of the curves.

$y = (x - 2)(x + 2)$

$y = (x - 1)(x - 3)$

$y = (x + 1)(x + 3)$

$y = -(x + 1)(x + 3)$

$y = (x + 2)(2 - x)$

$y = (x - 1)(x + 3)$

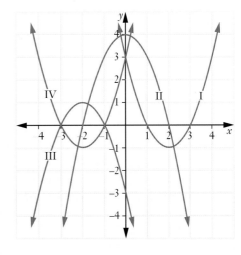

10 Given that each of the following tables show linear relationships between $x$ and $y$ find the rule for each table and complete each table.

**a**

| $x$ | 1 | 2 | 3 | 4 | 5 | 6 | 7 | 8 |
|---|---|---|---|---|---|---|---|---|
| $y$ | | | | 16 | 19 | | | |

**b**

| $x$ | 1 | 2 | 3 | 4 | 5 | 6 | 7 | 8 |
|---|---|---|---|---|---|---|---|---|
| $y$ | | | | | | | 13 | 15 |

**c**

| $x$ | 1 | 2 | 3 | 4 | 5 | 6 | 7 | 8 |
|---|---|---|---|---|---|---|---|---|
| $y$ | | | 11 | 9 | | | | |

**d**

| $x$ | 1 | 2 | 3 | 4 | 5 | 6 | 7 | 8 |
|---|---|---|---|---|---|---|---|---|
| $y$ | | 9 | | 19 | | | | |

**e**

| $x$ | 3 | 8 | 1 | 6 | 7 | 4 | 5 | 2 |
|---|---|---|---|---|---|---|---|---|
| $y$ | | 1 | | | | | 13 | |

11 Find the equation of the quadratic function that has exactly the same shape as that of $y = 3x^2$, has the same line of symmetry as $y = (x - 2)^2 + 3$ and that cuts the $y$-axis at the point (0, 15).

12 The diagram on the right shows a road under a bridge.

With $x$- and $y$-axes as shown the bridge arch (see diagram) has equation

$$y = \frac{5x}{16}(8 - x).$$

Determine

**a** the width of the road

**b** the clearance at the centre of the bridge.

For a vehicle of width 2.3 metres determine the maximum height of the vehicle (in whole centimetres) if it is to pass under the bridge and

**c** only use its own side of the road

**d** use both carriageways.

13 A stained glass window is to be constructed to the dimensions shown on the right. The arc BC is part of a circle which has the midpoint of AD as its centre and ABCD is rectangular. The window consists of four pieces of glass, I, II, III and IV separated and surrounded by strips of lead (shown as the straight and curved lines in the diagram).

Find

**a** the area of each piece of glass to the nearest square centimetre, (ignore the thickness of the lead strips),

**b** the total length of all the lead strips, to the nearest centimetre.

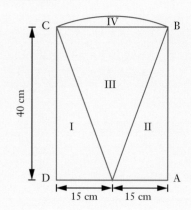

ISBN 9780170390330

# Quadratic equations

- What if the quadratic equation is not readily factorisable?
- Using the built-in facility of some calculators to solve equations
- Using a graphical approach
- Completing the square
- Obtaining and using a formula
- Miscellaneous exercise six

Try to solve the following problems. You might try to set up an equation and then solve it. If you find that approach difficult an alternative approach would be to try a value, test to see if it works and, if it does not, adjust your values for a new trial, i.e. 'trial and adjustment'.

## Situation One

I think of a number, multiply it by itself, take away four times the number I first thought of and end up with an answer of 21. What was the number I first thought of?

## Situation Two

I think of a number, multiply it by itself, take away ten times the number I first thought of, add 21 and end up with zero. What was the number I first thought of?

## Situation Three

If an object is dropped, the distance, $s$ metres, that it has fallen $x$ seconds later is given by the rule:

$$s = \frac{49}{10} x^2.$$

If an object is dropped from a bridge, how long does it take to reach the water, 65 metres below? (Give your answer to the nearest tenth of a second.)

## Situation Four

If an object is thrown vertically downwards with initial speed 5 m/s the distance, $s$ metres, that it has fallen $x$ seconds later is given by the rule:

$$s = 5x + \frac{49}{10} x^2.$$

If the object referred to in situation three were to be thrown vertically downwards with initial speed 5 m/s rather than dropped, how long does it take to reach the water now? (Again give your answer to the nearest tenth of a second.)

iStock.com/David Hughes

How did you get on with the situations on the previous page?

Perhaps for situation one you used trial and adjustment as shown below.

| | | | |
|---|---|---|---|
| Try $x = 1$, | $(1)^2 - 4(1) = -3$, | not 21. | Thus $x \neq 1$. |
| Try $x = 4$, | $(4)^2 - 4(4) = 0$, | not 21. | Thus $x \neq 4$. |
| Try $x = 6$, | $(6)^2 - 4(6) = 12$, | not 21. | Thus $x \neq 6$. |
| Try $x = 7$, | $(7)^2 - 4(7) = 21$, | as required. | Thus $x = 7$. |

The number first thought of was 7.

However, did you also discover that the number first thought of could have been $-3$?:

$$\text{If } x = -3, \qquad (-3)^2 - 4(-3) = 9 + 12$$
$$= 21, \qquad \text{as required.} \qquad \text{Thus } x = -3.$$

Alternatively, perhaps you set up an equation as shown below:

| | |
|---|---|
| Let $x$ be the number thought of: | $x$ |
| Multiply it by itself: | $x^2$ |
| Take away 4 times the number first thought of: | $x^2 - 4x$ |
| End up with 21: | $x^2 - 4x = 21$ |

The *Preliminary work* section at the beginning of this text mentioned that an ability to solve readily factorisable quadratic equations is assumed so you may have solved this equation as shown below, and arrived at the two possible solutions directly.

| | | | |
|---|---|---|---|
| Given | $x^2 - 4x = 21$, | | |
| i.e. | $x^2 - 4x - 21 = 0$ | | |
| Factorising | $(x - 7)(x + 3) = 0$ | | |
| Hence either | $x - 7 = 0$ | or | $x + 3 = 0$, |
| giving solutions of | $x = 7$, | | $x = -3$. |

Similarly did you discover the two possible numbers for situation two? In that case the equation was $x^2 - 10x + 21 = 0$ and both 3 and 7 satisfy this equation.

Situation three involved solving the equation
$$65 = \frac{49}{10}x^2,$$
$$\text{i.e.} \quad 650 = 49x^2,$$
$$\text{giving} \quad x^2 = \frac{650}{49}.$$

Once again there are two solutions to this equation. They are, correct to one decimal place, 3.6 and $-3.6$. However, given the context of the question, $x$ cannot be negative so there is only one realistic solution. Thus the object would take 3.6 seconds to hit the water.

Situation four involved solving the equation
$$65 = 5x + \frac{49}{10}x^2,$$
$$\text{i.e.} \quad 0 = 4.9x^2 + 5x - 65.$$

Once again there are two solutions to this equation, $x = 3.2$ and $x = -4.2$, correct to one decimal place, but only the positive answer is realistic, given the context. Thus the object would take 3.2 seconds to hit the water.

ISBN 9780170390330

All four of the situations involved solving equations that could be written in the form: $ax^2 + bx + c = 0$.

In situation one           $x^2 - 4x - 21 = 0$,    i.e.     $a = 1$,      $b = -4$,     $c = -21$.

In situation two           $x^2 - 10x + 21 = 0$,    i.e.     $a = 1$,      $b = -10$,    $c = 21$.

In situation three         $49x^2 - 650 = 0$,    i.e.     $a = 49$,    $b = 0$,       $c = -650$.

In situation four        $4.9x^2 + 5x - 65 = 0$,    i.e.     $a = 4.9$,    $b = 5$,      $c = -65$.

Equations that can be written in the form

$$ax^2 + bx + c = 0, a \neq 0$$

are called **quadratic** equations.

- In the previous chapter, we looked at the graphs of functions with a rule that could be written in the form $y = ax^2 + bx + c$, i.e. **quadratic functions**. For each value of $x$, the rule assigns one and only one value for $y$. Hence, a function.

- In this chapter we will consider ways of solving equations of the form $ax^2 + bx + c = 0$, i.e. **quadratic equations**. Solving the equation means finding the value(s) of $x$ for which the equality is true.

- Again, to repeat what was said in the *Preliminary work* section, it is anticipated that the reader is already familiar with solving 'readily factorisable quadratic equations'. In case this assumed ability is a little 'rusty', **Example 1** below, and the exercise that follows, provides practice.

- Solving a quadratic equation by factorising uses the fact that **if two numbers have a product of zero, then at least one of the numbers must be zero**. Just pause a moment to check that you understand and agree with this fact.
  Thus if $(x + 3)(x + 2) = 0$, it follows that either $(x + 3) = 0$ or $(x + 2) = 0$.

## EXAMPLE 1

Solve each of the following equations.

**a**   $x^2 - 6x = 0$             **b**   $x^2 - 2x - 3 = 0$          **c**   $x^2 - 13x = 30$

**d**   $x^2 = 15x$              **e**   $(2x - 5)(x + 7) = 0$       **f**   $x^2 = 16$

**g**   $2x^2 + 7x - 15 = 0$

### Solution

**a**   Given:               $x^2 - 6x = 0$

     Factorising:        $x(x - 6) = 0$

     Hence either       $x = 0$     or     $x - 6 = 0$,

     so                   $x = 0$     or      $x = 6$.

**b**   Given:               $x^2 - 2x - 3 = 0$

     (To factorise $x^2 - 2x - 3$ we look for two numbers that add to give –2 and multiply to give –3. The numbers are –3 and +1.)

     Factorising:      $(x - 3)(x + 1) = 0$

     Hence either       $x - 3 = 0$     or     $x + 1 = 0$,

     so                   $x = 3$     or      $x = -1$.

**c**  Given: $\qquad x^2 - 13x = 30,$

∴ $\qquad\qquad x^2 - 13x - 30 = 0$

(To factorise $x^2 - 13x - 30$ we look for two numbers that add to give –13 and multiply to give –30. The numbers are –15 and +2.)

Factorising: $\qquad (x - 15)(x + 2) = 0$

Either $\qquad\qquad x - 15 = 0 \qquad$ or $\qquad x + 2 = 0,$

so $\qquad\qquad\qquad x = 15 \qquad$ or $\qquad\qquad x = -2.$

**d**  Given $\qquad\qquad x^2 = 15x,$

∴ $\qquad\qquad x^2 - 15x = 0$

Factorising: $\qquad x(x - 15) = 0$

Either $\qquad\qquad x = 0 \qquad$ or $\quad x - 15 = 0,$

so $\qquad\qquad\qquad x = 0 \qquad$ or $\qquad\qquad x = 15.$

**e**  Given $\qquad (2x - 5)(x + 7) = 0 \qquad$ (Note that this is already in factorised form.)

Either $\qquad\qquad 2x - 5 = 0 \qquad$ or $\quad x + 7 = 0$

$\qquad\qquad\qquad\qquad 2x = 5 \qquad$ or $\qquad\qquad x = -7$

$\qquad\qquad\qquad\qquad x = 2.5 \qquad$ or $\qquad\qquad x = -7$

**f**  Given $\qquad\qquad x^2 = 16,$

$\qquad\qquad\qquad\qquad x = -4 \qquad$ or $\qquad\qquad +4,$

written $\qquad\qquad\qquad x = \pm 4.$

These answers could be obtained by factorising but it would be a longer process:

$$x^2 = 16$$
$$x^2 - 16 = 0$$
$$(x + 4)(x - 4) = 0$$

Either $\qquad\qquad x + 4 = 0 \qquad$ or $\quad x - 4 = 0,$

so $\qquad\qquad\qquad x = -4 \qquad$ or $\qquad\qquad x = 4.$

**g**  Given: $\qquad 2x^2 + 7x - 15 = 0$

(To factorise $ax^2 + bx + c$, when $a \neq 1$, we look for two numbers that add to give $b$ and multiply to give $ac$. We then rewrite $bx$ using these two numbers. Hence in this example we look for two numbers that add to give 7 and multiply to give –30. The numbers are 10 and –3.)

$$2x^2 + 7x - 15 = 0$$
$$2x^2 + 10x - 3x - 15 = 0$$
$$2x(x + 5) - 3(x + 5) = 0$$

∴ $\qquad\qquad (x + 5)(2x - 3) = 0$

Hence either $\qquad\qquad x + 5 = 0 \qquad$ or $\quad 2x - 3 = 0,$

so $\qquad\qquad\qquad x = -5 \qquad$ or $\qquad\qquad x = 1.5$

ISBN 9780170390330

**EXAMPLE 2**

I think of a number, multiply it by itself, take away six times the number I first thought of and end up with fifty five. What was the number I first thought of?

**Solution**

Let the number first thought of be $x$.

The given information leads to the equation $\quad x^2 - 6x = 55,$

i.e. $\qquad x^2 - 6x - 55 = 0$

$\qquad (x - 11)(x + 5) = 0$

Either $\qquad x - 11 = 0 \qquad$ or $\qquad x + 5 = 0,$

so $\qquad\qquad x = 11 \qquad$ or $\qquad x = -5.$

The number first thought of was either 11 or –5.


## Exercise 6A

Solve each of the following equations. (Without the assistance of your calculator.)

**1** $(x + 5)(x - 3) = 0$

**2** $(x + 8)(x + 9) = 0$

**3** $(2x - 11)(x + 5) = 0$

**4** $x^2 = 25$

**5** $x^2 - 49 = 0$

**6** $2x^2 = 200$

**7** $x^2 + 9x + 20 = 0$

**8** $x^2 + x - 20 = 0$

**9** $x^2 - 9x + 20 = 0$

**10** $x^2 - x - 20 = 0$

**11** $x^2 + 2x - 35 = 0$

**12** $x^2 + 4x + 3 = 0$

**13** $x^2 + 7x + 6 = 0$

**14** $x^2 + 10x + 21 = 0$

**15** $x^2 + 8x + 15 = 0$

**16** $x^2 - 4x - 12 = 0$

**17** $x^2 - 4x - 5 = 0$

**18** $x^2 - 4x = 0$

**19** $x^2 + 5x - 14 = 0$

**20** $x^2 - 36 = 0$

**21** $x^2 + 6x + 9 = 0$

**22** $x^2 - 3x - 4 = 0$

**23** $x^2 - 8x + 16 = 0$

**24** $x^2 = 15 - 2x$

**25** $x^2 = 3x$

**26** $x^2 + 12 = 7x$

**27** $x^2 = 24 - 10x$

**28** $4x^2 - 9 = 0$

**29** $25x^2 - 1 = 0$

**30** $x^2 = 2x + 15$

**31** $x^2 + 9 = 6x$

**32** $x^2 = 5(2x - 5)$

**33** $2x^2 + 5x - 12 = 0$

**34** $3x^2 + 10x - 8 = 0$

**35** $2x^2 - 3x - 5 = 0$

**36** $5x^2 + 34x - 7 = 0$

**37** $2x^2 + x - 21 = 0$

**38** $6x^2 - 19x + 10 = 0$

**39** $10x^2 - 9x + 2 = 0$

**40** When seven times a number is added to the square of the number the answer is 30. What is the number?

**41** When ten times a number is added to the square of the number and twenty five is added to the total the final answer is zero. What is the number?

**42** When an object is projected upwards from ground level, with initial speed 40 m/s, the height it is above ground $t$ seconds later is $h$ metres where $h$ is given by:

$$h \approx 40t - 5t^2.$$

State the value of $h$ when the object hits the ground again and find the value of $t$ then.

**43** If $s = ut + \dfrac{1}{2}at^2$ find the value of $t$ given that $s = 10$, $u = 3$, $a = 2$ and $t \geq 0$.

**44** If $w = kp^2 - 2cp$ find the value of $p$ given that $w = 33$, $k = 1$ and $c = 4$.

# What if the quadratic equation is not readily factorisable?

Given a quadratic equation $ax^2 + bx + c = 0$, when $ax^2 + bx + c$ is *not* readily factorised, we could

- use the built-in facility of some calculators to solve equations.

- use our ability to sketch the graph of the function $y = ax^2 + bx + c$ and then, from the sketch, estimate the values of $x$ for which $y = 0$.

- use the technique of completing the square from the previous chapter.

- use the technique of completing the square from the previous chapter to develop a formula for solving quadratic equations.

These approaches are discussed below.

## Using the built-in facility of some calculators to solve equations

Many calculators have built-in routines for solving quadratic equations.

In some we can type in the equation as it is, see below left, and in others we input the values of $a$, $b$ and $c$ and the calculator provides the solutions, see below right.

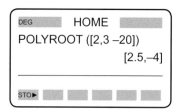

solve($2x^2 + 3x - 20 = 0, x$)
  {x = -4, x= 2.5}

DEG     HOME
POLYROOT ([2,3 -20])
              [2.5,-4]

STO▶

## Using a graphical approach

The characteristic shape of quadratic functions encountered in the last chapter explains why an equation of the form

$$ax^2 + bx + c = 0$$

can have more than one solution (also called *roots*), as explained below.

- If the graph of $y = ax^2 + bx + c$ cuts the $x$-axis in two places then there must be two places on the graph where $y = 0$, and hence two places where $ax^2 + bx + c = 0$.

  In such situations the quadratic equation $ax^2 + bx + c = 0$ will have *two* solutions.

- If the graph of $y = ax^2 + bx + c$ just touches the $x$-axis in one place then there must be just one place where $ax^2 + bx + c = 0$.

  In such situations the quadratic equation $ax^2 + bx + c = 0$ will have *one* solution.

- If the graph of $y = ax^2 + bx + c$ neither cuts nor touches the $x$-axis then there are no places where $ax^2 + bx + c = 0$.

  In such situations the quadratic equation $ax^2 + bx + c = 0$ will have *no* solutions.

ISBN 9780170390330

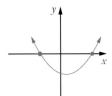

$y = ax^2 + bx + c$
cuts $x$-axis in two places.

$ax^2 + bx + c = 0$
has two solutions.

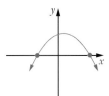

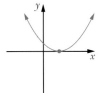

$y = ax^2 + bx + c$
touches $x$-axis in one place.

$ax^2 + bx + c = 0$
has one solution.

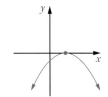

$y = ax^2 + bx + c$
neither cuts nor touches the $x$-axis.

$ax^2 + bx + c = 0$
has no solutions.

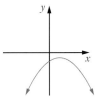

Thus to solve the equation $ax^2 + bx + c = 0$ we can sketch the graph of $y = ax^2 + bx + c$ and look at the $x$-coordinates of any points where the graph cuts the $x$-axis. At all such points $y$ will equal zero and hence $0 = ax^2 + bx + c$.

## EXAMPLE 3

Solve $2x^2 - 4x - 3 = 0$

**Solution**

Consider the function    $y = 2x^2 - 4x - 3$

Line of symmetry is      $x = 1$

Minimum point at         $(1, -5)$

Cuts $y$-axis at           $(0, -3)$

A sketch can then be made as shown on the right.

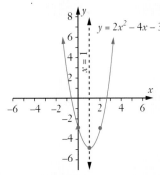

From this sketch:   $y = 0$   when   $x \approx -0.6$
                    and   when   $x \approx 2.6$

Hence $2x^2 - 4x - 3 = 0$ has solutions of   $x \approx -0.6$
                    and   $x \approx 2.6$.

Alternatively we could simply use a graphic calculator to display the graph of $y = 2x^2 - 4x - 3$ and locate the $x$-axis intercepts from the display.

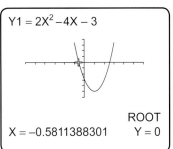

This gives solutions of $x = -0.58$ and $x = 2.58$ (2 decimal places).

# Completing the square

Completing the square

## EXAMPLE 4

Solve $x^2 - 4x - 1 = 0$

**Solution**

Given $$x^2 - 4x - 1 = 0$$

Create the gap: $$x^2 - 4x \qquad = 1$$

Insert the square of half the coefficient of $x$: $$x^2 - 4x + \left(\frac{-4}{2}\right)^2 = 1 + \left(\frac{-4}{2}\right)^2$$

Thus $$x^2 - 4x + 4 = 5$$
$$(x - 2)^2 = 5$$
$$x - 2 = \pm\sqrt{5}$$
$$x = 2 \pm \sqrt{5}$$

Giving solutions of 4.24 and –0.24 (2 decimal places).

## EXAMPLE 5

Solve $2x^2 + 14x - 5 = 0$

**Solution**

Given $$2x^2 + 14x - 5 = 0$$

Create the gap (and divide by 2): $$x^2 + 7x \qquad = \frac{5}{2}$$

Insert the square of half the coefficient of $x$: $$x^2 + 7x + \left(\frac{7}{2}\right)^2 = \frac{5}{2} + \left(\frac{7}{2}\right)^2$$

Thus $$\left(x + \frac{7}{2}\right)^2 = 14.75$$
$$x + 3.5 = \pm\sqrt{14.75}$$
$$x = -3.5 \pm \sqrt{14.75}$$

Giving solutions of 0.34 and –7.34 (2 decimal places).

ISBN 9780170390330

# Obtaining and using a formula

Given
$$ax^2 + bx + c = 0$$

Subtract $c$ from each side and then divide by $a$:
$$x^2 + \frac{b}{a}x = -\frac{c}{a}$$

Insert the square of half the coefficient of $x$:
$$x^2 + \frac{b}{a}x + \left(\frac{b}{2a}\right)^2 = -\frac{c}{a} + \left(\frac{b}{2a}\right)^2$$

Thus
$$\left(x + \frac{b}{2a}\right)^2 = -\frac{c}{a} + \frac{b^2}{4a^2}$$

$$= \frac{b^2 - 4ac}{4a^2}$$

$\therefore$
$$x + \frac{b}{2a} = \pm\sqrt{\frac{b^2 - 4ac}{4a^2}}$$

and so
$$x = -\frac{b}{2a} \pm \frac{\sqrt{b^2 - 4ac}}{2a}$$

$$= \frac{-b \pm \sqrt{b^2 - 4ac}}{2a}$$

If $ax^2 + bx + c = 0$, then $x = \dfrac{-b \pm \sqrt{b^2 - 4ac}}{2a}$

## EXAMPLE 6

Solve $2x^2 + 3x - 1 = 0$

**Solution**

Comparing $2x^2 + 3x - 1 = 0$ with $ax^2 + bx + c = 0$ gives $a = 2$, $b = 3$ and $c = -1$.

Substituting these values into the formula gives $x = \dfrac{-3 \pm \sqrt{3^2 - 4 \times 2 \times (-1)}}{2 \times 2}$

$$= \frac{-3 \pm \sqrt{17}}{4}$$

$$x = 0.28 \text{ or } x = -1.78 \text{ (2 decimal places)}.$$

Note:

- In the real number system, we cannot find the square root of a negative number. Hence if we are attempting to solve a quadratic $ax^2 + bx + c = 0$ for which the quantity $b^2 - 4ac$ is negative it is clear from the quadratic formula that we will have no real solutions.

  For example consider the quadratic equation $2x^2 + x + 2 = 0$.

  Comparing this equation with $ax^2 + bx + c = 0$ gives $a = 2$, $b = 1$ and $c = 2$.

  Thus $b^2 - 4ac = 1^2 - 4 \times 2 \times 2$
  $$= -15.$$

  The quadratic equation $2x^2 + x + 2 = 0$ will have no real solutions.

  This is also confirmed by the fact that the graph of $y = 2x^2 + x + 2$, shown on the right, does not cut the $x$-axis. I.e. there are no points on the graph for which $y = 0$.

  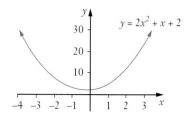

  However, if asked to solve $2x^2 + x + 2 = 0$, a calculator with a quadratic-solving program might not say 'no solutions' because some branches of mathematics use what are called complex numbers, and the equation does have solutions in this system. However, such work is beyond the requirement of this unit. We only need to know that a calculator response like that shown on the right should be interpreted as *no solutions*, or to be more correct, *no real solutions*. On some calculators a response like that shown can be avoided by setting the calculator to show real solutions only.

  $2x^2 + x + 2 = 0$

  | | x |
  |---|---|
  | 1 | −0.25 + 0.9682i |
  | 2 | −0.25 − 0.9682i |

  −0.25 + 0.9682458366i

- The quantity $b^2 - 4ac$ is called the **discriminant** of the quadratic equation $ax^2 + bx + c = 0$. It allows us to *discriminate* between (i.e. recognise the distinction between) the three situations of a quadratic having    *zero*,

            *one*

  or   *two*   real solutions.

  If $b^2 - 4ac > 0$, the quadratic equation $ax^2 + bx + c = 0$ will have two real solutions.

  If $b^2 - 4ac = 0$, the quadratic equation $ax^2 + bx + c = 0$ will have one real solution. (Sometimes referred to as one *repeated* root.)

  If $b^2 - 4ac < 0$, the quadratic equation $ax^2 + bx + c = 0$ will have no real solutions.

iStock.com/Andrew Penner

## Exercise 6B

### Using the built-in facility of some calculators to solve equations

Solving algebraic
equations

Use a calculator with a built-in routine for solving quadratic equations to solve questions 1 to 8, giving your answers correct to 2 decimal places.

**1** $3x^2 + x - 1 = 0$

**2** $x^2 + x - 3 = 0$

**3** $3x^2 + 3x + 1 = 0$

**4** $2x^2 + 6x + 1 = 0$

**5** $5x^2 + 7x - 3 = 0$

**6** $5x^2 + 6x = 2$

**7** $s = ut + \dfrac{1}{2}at^2$.

Find $t$ given that $s = 35$, $u = -25$, $a = 4$ and $t \geq 0$. (Give your answer correct to one decimal place.)

**8** $w = kp^2 - 2cp$.

Find $p$ given that $w = -3$, $k = 5$, $c = 7.5$. (Give your answers correct to two decimal places.)

### Using a graphical approach

Each of the following graphs are for equations of the form $y = ax^2 + bx + c$.
In each case state the number of real solutions $ax^2 + bx + c = 0$ appears to have.

**9**
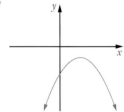

**10**
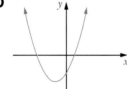

**11**
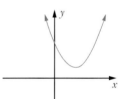

**12**
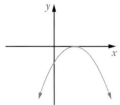

**13**
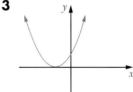

**14**
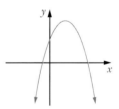

Each of the curves below have equations of the form $y = ax^2 + bx + c$.
In each case state the number of real solutions $ax^2 + bx + c = d$ appears to have.

**15**
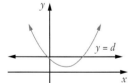

**16**
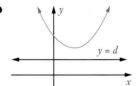

**17**
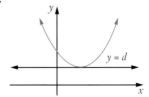

---

For each of the following, first sketch the graph of a suitable quadratic function and then use your sketch to solve the equation.

**18** $x^2 + 2x - 2 = 0$       **19** $x^2 + 4x - 7 = 0$       **20** $2x^2 - 8x + 3 = 0$

**21** $x^2 + 2x + 3 = 0$       **22** $2x^2 + 12x + 3 = 0$       **23** $1 + 4x - x^2 = 0$

## Completing the square

Solve each of the following quadratic equations using the technique of *completing the square*, giving your answers correct to two decimal places if rounding is necessary.

**24** $x^2 - 12x + 21 = 0$       **25** $x^2 - 6x + 10 = 0$       **26** $x^2 - 8x + 1 = 0$

**27** $x^2 + 7x - 5 = 0$       **28** $x^2 + 3x - 5 = 0$       **29** $2x^2 + x - 3 = 0$

Solve each of the following quadratic equations using the technique of *completing the square*, leaving your answers in the form $? \pm \sqrt{?}$, simplified if possible.

**30** $x^2 - 2x - 5 = 0$       **31** $x^2 - 6x + 1 = 0$       **32** $x^2 + 10x - 7 = 0$

**33** $2x^2 + 10x - 5 = 0$       **34** $3x^2 + 5x + 1 = 0$       **35** $5x^2 + x - 1 = 0$

## Using the formula

Solve each of the following quadratic equations using the quadratic formula:

$$x = \frac{-b \pm \sqrt{b^2 - 4ac}}{2a},$$

giving your answers correct to two decimal places.

**36** $x^2 + x - 4 = 0$       **37** $7x + 5 - 2x^2 = 0$       **38** $3x^2 + 1 = 7x$

**39** $6x = x^2 + 7$       **40** $x(x - 1) = 7$       **41** $2x(3x + 1) = 5$

Solve each of the following quadratic equations using the quadratic formula:

$$x = \frac{-b \pm \sqrt{b^2 - 4ac}}{2a},$$

leaving your answers in the form $? \pm \sqrt{?}$, simplified if possible.

**42** $x^2 + 3x + 1 = 0$       **43** $x^2 - 7x + 1 = 0$       **44** $2x^2 + x - 5 = 0$

**45** $3x^2 = 1 + 5x$       **46** $5x^2 - 5 + x = 0$       **47** $2x(x + 2) = -1$

By determining the discriminant of each of the following quadratic equations, determine the number of real roots each equation has (and then check your answers using the equation-solving ability of some calculators).

**48** $x^2 + 5x - 7 = 0$       **49** $x^2 + 5x + 7 = 0$       **50** $x^2 - 2x - 3 = 0$

**51** $2x^2 + 7x + 5 = 0$       **52** $4x^2 - 12x + 9 = 0$       **53** $3x^2 - x + 1 = 0$

# Miscellaneous exercise six

This miscellaneous exercise may include questions involving the work of this chapter, the work of any previous chapters, and the ideas mentioned in the Preliminary work section at the beginning of the book.

**1** The product of 3 less than a number and 5 more than the number is zero. What could the number be?

**2** The display on the right shows the lines

$x = 60$,            $y = 60$,

$y = 2x - 60$,      $y = 0.5x + 30$,

$y = -x + 60$,      $y = -2x + 30$

labelled A to F.

Allocate the correct rule to each line.

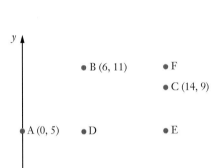

**3** In the diagram on the right the points A, D and E form a horizontal line, as do points B and F. Similarly points E, C and F form a vertical line, as do points D and B.

With the coordinates of A, B and C as indicated in the diagram determine:

  **a** the length of AD, the length of DB and hence the gradient of the straight line through A and B.

  **b** the length of DE, the length of EC and hence the gradient of the straight line through D and C.

  **c** the gradient of the straight line through D and F.

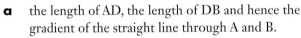

**4** The graph of the quadratic function $y = 3(x - 1)^2 + 2$ has a line of symmetry with equation $x = a$, has a minimum point at $(b, c)$ and passes through the points $(6, d)$, $(-4, e)$ and $(f, 14)$. Determine $a, b, c, d, e$ and the two possible values of $f$.

**5** A rectangle is such that multiplying the width by five gives an answer that is three centimetres less than the length of the rectangle. If the area of the rectangle is 36 cm$^2$ find the dimensions of the rectangle.

**6** OAB is a right triangle with OA = 5 cm, AB = 10 cm and ∠OAB = 90°. A circle of radius 5 cm is drawn, centre O. Find the area of that part of triangle OAB not lying in the circle, giving your answer in square centimetres correct to one decimal place.

**7** Each of the curves A to H shown below are for functions of the form
$$y = a(x - p)^2 + q$$
for integer $a$, $p$ and $q$. Determine the equation of each curve.

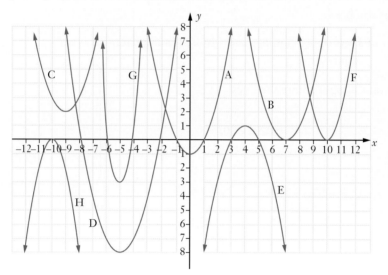

**8** A solid cylinder is to be made having a length of 30 cm and a total surface area of 2000 cm². 

Determine the radius of such a cylinder, to the nearest millimetre, clearly showing your use of the quadratic formula in your working.

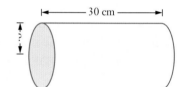

**9** Determine the value of $x$ in the right triangle shown sketched on the right, giving your answer rounded to two decimal places and clearly showing your use of the technique of *completing the square* in your working.

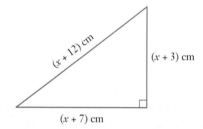

**10** Apply the technique of completing the square to the general quadratic equation
$$ax^2 + bx + c = 0$$
to obtain the quadratic formula.
$$x = \frac{-b \pm \sqrt{b^2 - 4ac}}{2a}.$$

ISBN 9780170390330

# 7.

# Polynomials and other functions

- Polynomial functions
- Cubic functions
- Cubics in factorised form
- Sketching cubic functions
- Transformations
- What will the graphs of $y = \dfrac{1}{x}$ and $y = \sqrt{x}$ look like?
- Vocabulary
- Transformations of the general function $y = f(x)$
- Two relationships that are not functions
- Miscellaneous exercise seven

# Polynomial functions

In a linear function the highest power of $x$ is 1: $\qquad y = mx^1 + c$

In a quadratic function the highest power of $x$ is 2: $\qquad y = ax^2 + bx + c$

Continuing this pattern we have *cubic* functions: $\qquad y = ax^3 + bx^2 + cx + d$

$\qquad\qquad\qquad$ *quartic* functions: $\qquad y = ax^4 + bx^3 + cx^2 + dx + e$

$\qquad\qquad\qquad\qquad\qquad\qquad$ etc.

These functions are all part of the larger family of *polynomial functions*.

Polynomial functions have rules of the form:

$$f(x) = a_n x^n + a_{n-1} x^{n-1} + a_{n-2} x^{n-2} + \ldots + a_2 x^2 + a_1 x + a_0$$

where $n$ is a non-negative integer and $a_n, a_{n-1}, a_{n-2}, \ldots$ are all numbers, called the **coefficients** of $x^n, x^{n-1}, x^{n-2}$ etc.

The highest power of $x$ is the **order** of the polynomial.

Thus linear functions, $y = mx + c$, are polynomials of order 1, quadratic functions, $y = ax^2 + bx + c$, are polynomials of order 2.

We will now consider **cubic** functions, i.e. polynomials of order 3.

# Cubic functions

With $a = 1$ and $b = c = d = 0$, the general formula for a cubic function

$$y = ax^3 + bx^2 + cx + d$$

reduces to $\qquad y = 1x^3 + 0x^2 + 0x + 0,$

i.e. the most basic cubic: $\quad y = x^3$.

A table of values and the graph of $y = x^3$ are shown below:

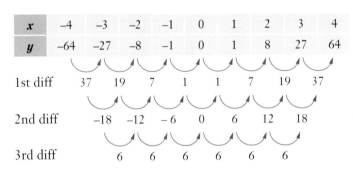

| $x$ | $-4$ | $-3$ | $-2$ | $-1$ | 0 | 1 | 2 | 3 | 4 |
|---|---|---|---|---|---|---|---|---|---|
| $y$ | $-64$ | $-27$ | $-8$ | $-1$ | 0 | 1 | 8 | 27 | 64 |

1st diff $\quad\quad$ 37 $\quad$ 19 $\quad$ 7 $\quad$ 1 $\quad$ 1 $\quad$ 7 $\quad$ 19 $\quad$ 37

2nd diff $\quad\quad\quad$ $-18$ $\quad$ $-12$ $\quad$ $-6$ $\quad$ 0 $\quad$ 6 $\quad$ 12 $\quad$ 18

3rd diff $\quad\quad\quad\quad$ 6 $\quad$ 6 $\quad$ 6 $\quad$ 6 $\quad$ 6 $\quad$ 6

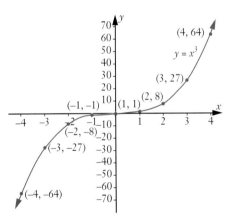

### Note

Note that the table of values has a constant third difference – a characteristic of the tables of values for cubic functions.

The graph of $y = x^3$ has certain features that, with thought, we should have expected:

- The cube of any positive number is positive and the cube of a negative number is negative. Thus we would only expect to find $y = x^3$ in the 1st quadrant (see diagram), as that is where both the $x$ and $y$ coordinates are positive, and in the 3rd quadrant, as that is where both $x$ and $y$ coordinates are negative.

  | | |
  |---|---|
  | 2nd Quadrant | 1st Quadrant |
  | 3rd Quadrant | 4th Quadrant |

- We would expect the graph to pass through $(0, 0)$.

- As $x$ gets large positively, we would expect $y$ to be even larger and positive.

- As $x$ gets large negatively, we would expect $y$ to be even larger and negative.

- For every point $(x, y)$ on the graph there will also exist a point $(-x, -y)$, e.g. $(2, 8)$ and $(-2, -8)$, $(3, 27)$ and $(-3, -27)$. I.e. for $f(x) = x^3$, $f(-a) = -f(a)$. This gives the graph its rotational symmetry.

These 'expected features' are indeed evident on the graph.

Cubic functions will not all give the same shape graph as that of $f(x) = x^3$ but they all

- have either two turning points or no turning points,

  e g:

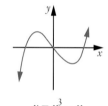 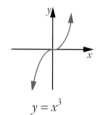 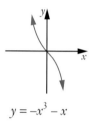

$y = x^3 - x$         $y = x^3$         $y = -x^3 - x$

- cut the $y$-axis once and cut (or touch) the $x$-axis in 1, 2 or 3 places:

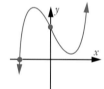 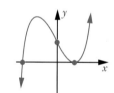 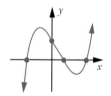

**EXAMPLE 1**

Determine the coordinates of the point where the graph of the cubic function

$$y = 2x^3 + 3x^2 + 2x + 14 \text{ cuts the } y\text{-axis.}$$

**Solution**

All points on the $y$-axis have an $x$-coordinate of zero.

If $x = 0$,            $y = 2(0)^3 + 3(0)^2 + 2(0) + 14$

                          $= 14$

The given cubic function cuts the $y$-axis at the point $(0, 14)$.

# Cubics in factorised form

Just as some quadratic functions, $y = ax^2 + bx + c$,
may be expressed in the factorised form, $y = a(x - p)(x - q)$,
similarly some cubic functions $y = ax^3 + bx^2 + cx + d$,
may be expressed in the factorised form $y = a(x - p)(x - q)(x - r)$.

Determine the coordinates of the points where the graph of the cubic function
$y = (x - 1)(2x - 3)(x + 2)$ cuts

**a**   the $x$-axis,

**b**   the $y$-axis.

### Solution

**a**   All points on the $x$-axis have a $y$-coordinate of zero.

If $y = 0$,    $0 = (x - 1)(2x - 3)(x + 2)$

If three brackets have a product of zero,
one of the brackets must equal zero.

i.e.    $(x - 1) = 0$   or   $(2x - 3) = 0$   or   $(x + 2) = 0$

giving         $x = 1$   or         $2x = 3$   or         $x = -2$

i.e.   $x = 1.5$

The graph cuts the $x$-axis at $(1, 0)$, $(1.5, 0)$ and $(-2, 0)$.

**b**   All points on the $y$-axis have an $x$-coordinate of zero.

Given         $y = (x - 1)(2x - 3)(x + 2)$

If $x = 0$,     $y = (0 - 1)(2 \times 0 - 3)(0 + 2)$

$= (-1)(-3)(2)$

$= 6$

The graph cuts the $y$-axis at $(0, 6)$.

The graph of $y = (x - 1)(2x - 3)(x + 2)$ is shown on the right.

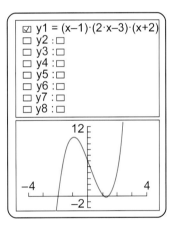

---

Use your graphic calculator to view the graphs of cubic functions of the form

$$y = (x - b)(x - c)^2$$

i.e.                $y = (x - b)(x - c)(x - c)$

and of the form

$$y = (x - b)^3$$

i.e.                $y = (x - b)(x - b)(x - b)$.

What effect does the presence of a 'repeated bracket' have on the graph of the function?

---

**7.** Polynomials and other functions ●●●●●●

# Sketching cubic functions

Nowadays, with the ready access to graphic calculators, the easiest way to view the graph of a cubic function is to display it on a calculator (as in **Example 4** on the next page). However, if you are required to make a sketch without using a graphic calculator, and provided the cubic is given in factorised form, or in a form that can be readily factorised, for example $y = x^3 - 5x^2 + 6x$, we can determine sufficient information about the cubic for a reasonable sketch to be made, as in **Examples 3** and **5** that follow. (Or, if not, other points lying on the curve can be determined for other $x$ values.)

### EXAMPLE 3

For the cubic function $y = -2(x + 3)(x - 2)(x - 5)$, find the coordinates of any points where the graph of the function cuts

**a**   the $y$-axis,   **b**   the $x$-axis.

Describe the behaviour of the $y$ values as the $x$ values become

**c**   increasingly large positively,   **d**   increasingly large negatively.

**e**   Hence sketch the function.

**Solution**

**a**   All points on the $y$-axis have an $x$-coordinate of zero.
   If $x = 0$,    $y = -2(0 + 3)(0 - 2)(0 - 5)$
                    $= -60$
   The graph cuts the $y$-axis at $(0, -60)$.

**b**   All points on the $x$-axis have a $y$-coordinate of zero.
   If $y = 0$,    $0 = -2(x + 3)(x - 2)(x - 5)$
   i.e. $x + 3 = 0$ or $x - 2 = 0$ or $x - 5 = 0$
   The graph cuts the $x$-axis at $(-3, 0)$, $(2, 0)$ and $(5, 0)$.

**c**   As $x$ becomes increasingly large positively (or negatively), its value will dominate each of the three brackets. Thus:
   For $x$ large and positive   $y = -2(x + 3)(x - 2)(x - 5)$
   becomes                          $y = -2 \times (\text{large +ve}) \times (\text{large +ve}) \times (\text{large +ve})$
                                        $= $ a very large negative number.
   As $x$ becomes increasingly large positively, $y$ becomes very large negatively.
   I.e.: As $x \to \infty$, $y \to -\infty$.

**d**   For $x$ large and negative   $y = -2(x + 3)(x - 2)(x - 5)$
   becomes                          $y = -2 \times (\text{large -ve}) \times (\text{large -ve}) \times (\text{large -ve})$
                                        $= $ a very large positive number.
   As $x$ becomes increasingly large negatively, $y$ becomes very large positively.
   i.e.: As $x \to -\infty$, $y \to \infty$.

**e**   Placing these facts on a graph, below left, allows a sketch to be made, below right.

ISBN 9780170390330

## EXAMPLE 4

Use a graphic calculator to view the graph of the cubic function

$$y = 2x^3 - 11x^2 + 17x - 13.$$

Use your calculator to determine the coordinates of the point(s) where the function cuts the $x$-axis, rounding any $x$-coordinates to two decimal places.

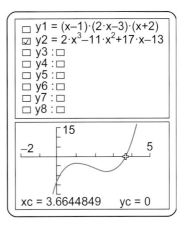

### Solution

A typical calculator display is shown on the right and includes the coordinates of the point where the graph cuts the $x$-axis.

Hence the coordinates are, to the required accuracy, $(3.66, 0)$.

## EXAMPLE 5

Without the assistance of a calculator produce a sketch of the graph of the cubic function with equation:

$$y = 2(x + 1)(x - 5)^2.$$

### Solution

If $y = 0$, $\qquad\qquad\qquad\qquad 0 = 2(x + 1)(x - 5)^2.$

Hence the cubic function cuts the $x$-axis at the point $(-1, 0)$ and *touches* the $x$-axis at the point $(5, 0)$ (the repeated bracket indicating the 'touch' of the $x$-axis as you may have discovered from the investigation on page 131).

If $x = 0$, $\qquad\qquad\qquad\qquad y = 2(0 + 1)(0 - 5)^2$
$$= 50.$$

Hence the cubic function cuts the $y$-axis at the point $(0, 50)$.

As $x$ gets large positively, $y$ gets very large positively. $\qquad$ (As $x \to \infty, y \to \infty$.)

As $x$ gets large negatively, $y$ gets very large negatively. $\qquad$ (As $x \to -\infty, y \to -\infty$.)

Placing these facts on a graph allows a sketch to be made as shown below (or, if still in doubt, locate another point, say when $x = 1$ to give the point $(1, 64)$).

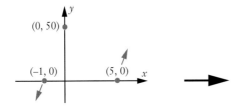 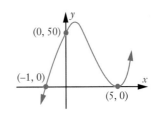

## Exercise 7A

**1** Determine the coordinates of the point where the graph of each of the following cubic functions cuts the $y$-axis.

**a** $y = x^3 + x^2 + x + 1$      **b** $y = 3x^3 - 5x^2 - 2x - 5$      **c** $y = x^3 + 8$

**d** $y = 2x^3 + 3x^2 + 6$      **e** $y = 2 + 3x + 7x^2 - x^3$      **f** $y = 5x + 3 + 2x^3$

**2** Determine the coordinates of the point(s) where the graph of each of the following cubic functions cut, or perhaps just 'touches', the $x$-axis.

**a** $y = (x - 2)(x - 3)(x - 4)$      **b** $y = (x + 7)(x - 1)(x - 5)$      **c** $y = (2x - 5)(x + 1)(5x - 3)$

**d** $y = (1 - x)(1 + x)(x - 7)$      **e** $y = x(4x - 1)(2x - 7)$      **f** $y = (x + 1)^2(x - 5)$

**g** $y = x^3 - 9x$      **h** $y = x^3 + 2x^2 - 15x$

**3** Use a graphic calculator to view the graph of the cubic function $y = 2x^3 - 2x^2 - 3x - 5$.

Use your calculator to determine the coordinates of the point(s) where the function cuts the $x$-axis, rounding any $x$-coordinate(s) to two decimal places.

**4 a** Given that $x^3 + 5x^2 - 12x - 36 = (x + 2)(x - 3)(x - k)$, find $k$.

**b** Find the coordinates of any point(s) where $y = x^3 + 5x^2 - 12x - 36$ cuts the $x$-axis.

**5** If $f(x) = x^3 - 6x^2 - x + 6$ determine

**a** $f(-1)$      **b** $f(1)$      **c** $f(2)$      **d** $f(6)$.

Hence factorise $x^3 - 6x^2 - x + 6$.

**6** If $f(x) = x^3 - 10x^2 + 31x - 30$ determine

**a** $f(1)$      **b** $f(2)$      **c** $f(3)$.

Hence factorise $x^3 - 10x^2 + 31x - 30$.

**7** Given that $3x^3 - 14x^2 - 7x + 10 = (3x - 2)(ax^2 + bx + c)$:

**a** Determine the value of $a$ and the value of $c$ by inspection.

**b** With your answers from part **a** in place expand $(3x - 2)(ax^2 + bx + c)$ and hence determine $b$.

**c** Find the coordinates of any $x$-axis intercepts of the graph of $y = 3x^3 - 14x^2 - 7x + 10$.

**8** By determining

- the coordinates of any points where the function cuts, or perhaps just touches the axes

- the behaviour of the function as $x$ gets large positively and negatively,

produce sketches of each of the following cubic functions. Then check the reasonableness of each sketch by viewing the graph of the function on a graphic calculator.

**a** $y = (x + 2)(x - 2)(x - 5)$      **b** $y = (x + 4)(x + 1)(x - 5)$      **c** $y = 2(x + 4)(x + 1)(x - 5)$

**d** $y = x(3 - x)(x - 7)$      **e** $y = (x - 1)(x - 3)^2$      **f** $y = (x - 2)^3$

# Transformations

From Chapter 5, *Quadratic functions*, you should be familiar with the following ideas:

- Altering the '$a$' in $y = ax^2$ stretches, *dilates*, the graph vertically. (Points on the $x$-axis are unmoved.)

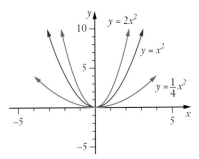

- The graphs of $y = -x^2$, $y = -2x^2$ etc. are simply those of $y = x^2$, $y = 2x^2$ etc., *reflected* in the $x$-axis.

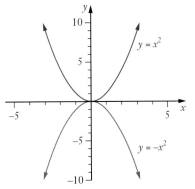

- Altering the '$c$' in $y = x^2 + c$ *translates* the graph vertically.

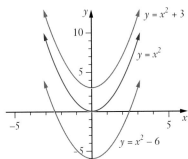

- Altering the '$b$' in $y = (x - b)^2$ *translates* the graph horizontally.

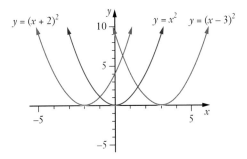

Hence the graph of $y = a(x - b)^2 + c$ is that of $y = x^2$ stretched vertically, scale factor $a$ (and reflected in the $x$-axis if $a$ is negative), translated $b$ units right and then $c$ units up.

## TECHNOLOGY INVESTIGATION

**Do such ideas also apply to cubic functions?**

Does altering the '$a$' in $y = ax^3$ stretch (dilate) the graph vertically?

Are the graphs of $y = -x^3$, $y = -2x^3$ etc. simply those of $y = x^3$, $y = 2x^3$ etc., reflected in the $x$-axis?

Does altering the '$c$' in $y = x^3 + c$ translate the graph vertically?

Does altering the '$b$' in $y = (x - b)^3$ translate the graph horizontally?

**So how does the graph of $y = a(x - b)^3 + c$ compare to the graph of $y = x^3$?**

# What will the graphs of $y = \dfrac{1}{x}$ and $y = \sqrt{x}$ look like?

$y = \dfrac{1}{x}$

The *Preliminary work* section at the beginning of this book reminded us that functions of the form $y = \dfrac{k}{x}$

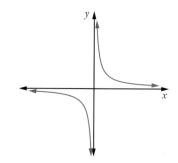

- have graphs with the characteristic shape shown on the right (reflected in the $y$-axis if $k$ is negative).

- describe situations in which the two variables are inversely proportional.

- have tables of values for which paired values have a common product.

Hence $y = \dfrac{1}{x}$ has this characteristic shape:

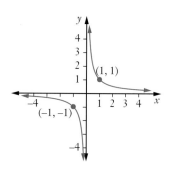

Again, with thought, this shape is exactly as we should have expected for $y = \dfrac{1}{x}$ because:

- When $x$ is positive, $y$ will be positive, and when $x$ is negative, $y$ will be negative. Thus the graph only exists where $x$ and $y$ are of the same sign.

- If $x$ is large, $y$ must be small, and if $x$ is small, $y$ must be large.

- The function does not exist for $x = 0$ and there are no values of $x$ for which $y$ equals 0. Thus the function does not cut either axis.

## Note

- The *x*-axis is said to be a horizontal **asymptote** to the curve and the *y*-axis is a vertical **asymptote**. These are lines that the curve gets closer and closer to without ever quite touching.

- For every point $(x, y)$ on the graph there will also exist a point $(-x, -y)$, for example $(1, 1)$ and $(-1, -1)$, $(2, 0.5)$ and $(-2, -0.5)$. I.e. $f(-a) = -f(a)$. This gives the graph its rotational symmetry.

- The graph is said to be **hyperbolic**.

ISBN 9780170390330

# $y = \sqrt{x}$

Again let us think about some of the characteristics we would expect the graph of $y = \sqrt{x}$ to possess.

- In our system of real numbers we cannot determine the square root of a negative number. Hence the graph does not exist for negative values of $x$.

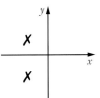

- There are also no values of $x$ for which $\sqrt{x}$ will be negative.

  Hence the graph does not exist for negative values of $y$.

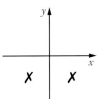

Putting the two previous ideas together we have $y = \sqrt{x}$ only existing where $x$ and $y$ both take non negative values.

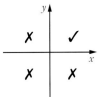

- We would expect the graph of $y = \sqrt{x}$ to include the point $(0, 0)$ and as $x$ gets large positively $y$ would also get large positively but at a slower rate.

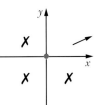

These 'expected features' are indeed evident on the graph of $y = \sqrt{x}$.

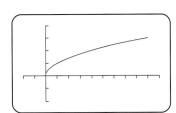

> **INVESTIGATION**
>
> How does the graph of $y = \dfrac{a}{x-b} + c$ compare to the graph of $y = \dfrac{1}{x}$?
>
> How does the graph of $y = a\sqrt{x-b} + c$ compare to the graph of $y = \sqrt{x}$?
>
> Investigate and write a summary of your findings.

# Vocabulary

Some vocabulary that we will commonly encounter when discussing the shape and notable features of the graphs of functions is explained below. Some of these terms we are already familiar with from this and previous chapters of this text.

For example, an earlier page considered a curve that got closer and closer to some lines without ever quite touching the lines and referred to these lines as **asymptotes**.

In the diagram below the lines $x = 3$ and $y = 2$ are asymptotes to the curve.

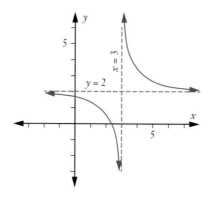

The graph shown on the right does not appear to have any asymptotes but some other noteworthy features are shown.

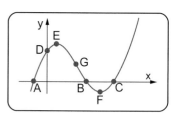

- The *x*-axis intercepts. Points A, B and C in the diagram.

- The *y*-axis intercepts. Point D in the diagram.

- Any **turning points** the graph has. Points E and F in the diagram.

  E is a **maximum turning point** and *F* is a **minimum turning point**.

- If $y = f(x)$ is shaped ∩ (or part of ∩) we say that it is **concave down**.

  The graph shown appears to be concave down to the left of point G.

- If $y = f(x)$ is shaped ∪ (or part of ∪) we say that it is **concave up**.

  The graph shown appears to be concave up to the right of point G.

- The points on a curve where it changes from being concave down to concave up, or from concave up to concave down, are called **points of inflection**. Point G in the above diagram is a point of inflection. If, at a point of inflection, the graph is momentarily horizontal then the point is a point of **horizontal inflection**.

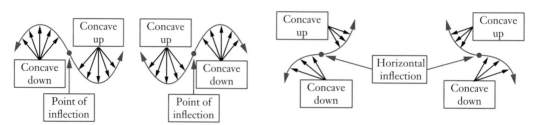

Also remember:

- Some graphs possess **symmetry**.

  Eg. $y = x^2$ has *line symmetry*, $y = x^3$ has *rotational symmetry*.

- Some functions have regions on the graph where the function is undefined. The natural domain is then not the entire set of real numbers. For example $y = \sqrt{x}$ is undefined for $x < 0$. The natural **domain** of $y = \sqrt{x}$ is $\{x \in \mathbb{R}: x \geq 0\}$.

- Some functions do not output all of the real numbers. For example $y = x^2 + 1$ will not output any numbers less that 1. Hence for $x \in \mathbb{R}$ the function $y = x^2 + 1$ has a **range** of $\{y \in \mathbb{R}: y \geq 1\}$.

## Exercise 7B

**1** Graph A shown in red on the right has equation $y = \sqrt{x}$.

Graphs B, C and D are all translations of graph A.

Write down the equations of B, C and D.

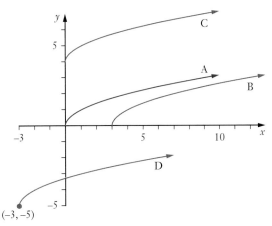

**2** Find the equation of each graph below, given each is of the form $y = \dfrac{1}{x} + c$.

**a**

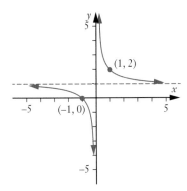

**b**

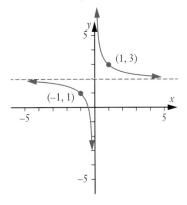

**c**

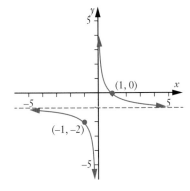

**7.** Polynomials and other functions

**3** Find the equation of each graph below, given each is of the form $y = \dfrac{1}{x+c}$.

**a**

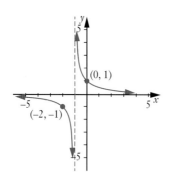

(0, 1)

(−2, −1)

**b**

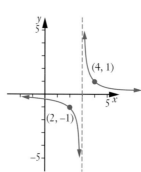

(4, 1)

(2, −1)

**c**

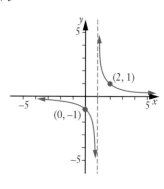

(2, 1)

(0, −1)

**4** Describe how the graph of $y = x^3 + 1$ compares to that of $y = x^3$.

**5** Describe how the graph of $y = \dfrac{1}{x-1}$ compares to that of $y = \dfrac{1}{x}$.

**6** Describe how the graph of $y = 2\sqrt{x}$ compares to that of $y = \sqrt{x}$.

**7** Describe how the graph of $y = (x-3)^2$ compares to that of $y = (x+4)^2$.

**8** Describe how the graph of $y = \sqrt{x-2} + 1$ compares to that of $y = \sqrt{x}$.

**9** Describe how the graph of $y = \dfrac{3}{x-1}$ compares to that of $y = \dfrac{1}{x}$.

**10** For the function graphed state which of the points A to I are:

**a**  the 2 maximum turning points.

**b**  the minimum turning point.

**c**  the four points of inflection.

**d**  the one point of horizontal inflection.

Between which points is the function

**e**  concave up?      **f**  concave down?

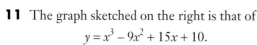

**11** The graph sketched on the right is that of

$$y = x^3 - 9x^2 + 15x + 10.$$

By viewing the graph of the function on a graphic calculator determine the coordinates of points A to G which are all the intercepts with the axes, the turning points and the point of inflection (which in this case is halfway between the two turning points).

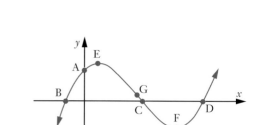

**12** According to Boyle's law, the pressure of a fixed mass of gas, kept at a constant temperature, is inversely proportional to the volume of the gas. i.e. $P = \dfrac{k}{V}$.

Let us suppose that for a particular, non zero, mass of a gas the pressure, $P$ units of pressure, and the volume, $V \text{ cm}^3$, are such that $P = \dfrac{400}{V}$.

**a** Find $V$ when $P = 40$.

**b** Find $V$ when $P = 20$.

**c** What would be a suitable domain for $V$?

**13** The nine equations given below are for the nine graphs shown below. Determine the coordinates of points A, B, C, … I and the values of $a, b, c, \ldots i$.

$y = x^3 + 8$

$y = (x + 1)^3 + 8$

$y = (x - d)^2 + e$

$y = \sqrt{x + a}$

$y = \dfrac{8}{x} - 2$

$y = \dfrac{1}{x - f} + g$

$y = x^2 + 4$

$y = b(x - c)^2$

$y = hx + i$

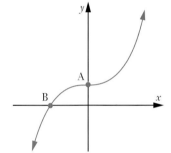

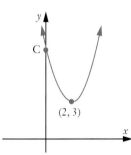

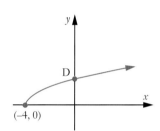

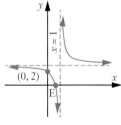

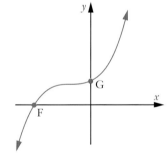

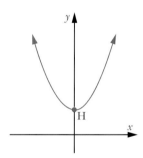

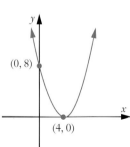

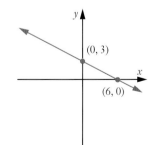

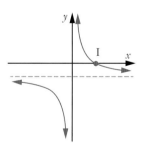

# Transformations of the general function y = f(x)

Graphing translations
of functions

Given the graph of $y = x^2$, or $y = x^3$, or $y = \sqrt{x}$ or $y = \dfrac{1}{x}$ we have seen that 'adding $k$ to the right hand side' of these equations moves the original graph up $k$ units. This, and other transformations given in terms of a more general function $y = f(x)$ are stated below. Note especially 'replacing $x$ by $-x$' and 'replacing $x$ by $ax$', because those transformations have not been encountered so far in this text.

To allow us to illustrate the various transformations let us suppose that some function $y = f(x)$ has the graph shown on the right. (There is no significance in the shape chosen. It could be the graph of any function.)

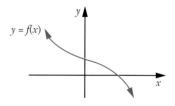

## 'Adding k to the right hand side'

The graph of $y = f(x) + k$ will be that of $y = f(x)$ translated $k$ units vertically upwards. Thus if $k$ is negative the translation will be vertically downwards.

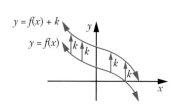

## 'Replacing x by (x − k)'

The graph of $y = f(x - k)$ will be that of $y = f(x)$ translated $k$ units to the right. Thus if $k$ is negative the translation will be to the left.

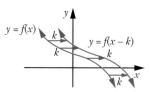

## 'Multiplying the right hand side by −1'

The graph of $y = -f(x)$ will be that of $y = f(x)$ reflected in the $x$-axis.

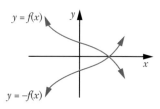

## 'Replacing x by −x'

The graph of $y = f(-x)$ will be that of $y = f(x)$ reflected in the $y$-axis.

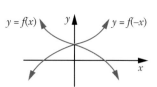

## 'Multiplying the right hand side by a'

The graph of $y = af(x)$ will be that of $y = f(x)$ dilated parallel to the $y$-axis with scale factor $a$. A point that is $q$ units above the $x$-axis will be moved vertically to a point that is $aq$ units above the $x$-axis. Points on the $x$-axis will not move.

If $a > 1$ the effect will be to stretch $y = f(x)$ vertically and if $0 < a < 1$ the effect will be to compress $y = f(x)$ vertically.

ISBN 9780170390330

Below left shows the situation for $a = 2$ and below right shows $a = 0.5$.

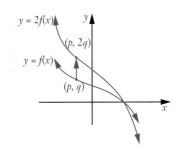 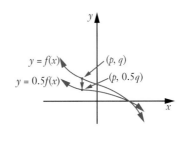

## 'Replacing x by ax'

The graph of $y = f(ax)$ will be that of $y = f(x)$ dilated parallel to the $x$-axis with scale factor $\dfrac{1}{a}$. A point that is $p$ units from the $y$-axis will be moved horizontally to a point that is $\dfrac{p}{a}$ units from the $y$-axis. Points on the $y$-axis will not move.

If $a > 1$ the effect will be to compress $y = f(x)$ horizontally and if $0 < a < 1$ the effect will be to stretch $y = f(x)$ horizontally.

Below left shows the situation for $a = 2$ and below right shows $a = 0.5$.

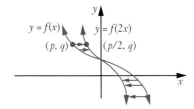 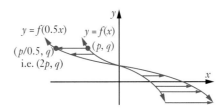

For 'half the $x$ value' $f(2x)$ will output the same value as $f(x)$ would.

We now need 'twice the $x$ value' for $f(0.5x)$ to output the same value as $f(x)$ would.

Using a viewing window of –6 to 6 on the $x$-axis and –8 to 8 on the $y$-axis display the graph of $y = x^3 + 2x^2 - x + 3$ on your graphic calculator.

For each of the equations I to VI given below:

- Predict what transformation of the graph of $y = x^3 + 2x^2 - x + 3$ will give the graph of the given equation.

- Display the graph of the given equation on your graphic calculator, along with that of $y = x^3 + 2x^2 - x + 3$, to test your prediction.

    I:    $y = -(x^3 + 2x^2 - x + 3)$
    II:   $y = (x - 2)^3 + 2(x - 2)^2 - (x - 2) + 3$
    III:  $y = (-x)^3 + 2(-x)^2 - (-x) + 3$
    IV:  $y = 0.5(x^3 + 2x^2 - x + 3)$
    V:   $y = (0.5x)^3 + 2(0.5x)^2 - (0.5x) + 3$
    VI:  $y = x^3 + 2x^2 - x - 2$

## EXAMPLE 6

The graph of $y = f(x)$ is shown.

The 'filled' and 'empty' circles indicate where the function is (filled circle) and is not (empty circle). Thus $f(0) = 0$, not $-2$.

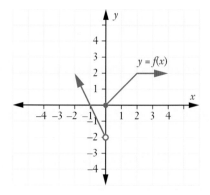

Draw the graph of each of the following.

a   $y = f(x) + 1$         b   $y = f(x + 2)$

c   $y = 2f(x)$           d   $y = f(0.5x)$

e   $y = f(2x)$           f   $y = -f(x)$

### Solution

a   To go from             $y = f(x)$
    to                     $y = f(x) + 1$
    involves adding 1 to the right hand side.

    Thus the graph of      $y = f(x) + 1$
    will be that of        $y = f(x)$
    translated vertically upwards 1 unit.

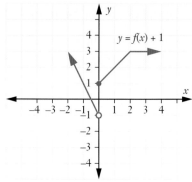

b   To go from             $y = f(x)$
    to                     $y = f(x + 2)$
    involves replacing $x$ by $x + 2$.

    Thus the graph of      $y = f(x + 2)$
    will be that of        $y = f(x)$
    translated 2 units to the left.

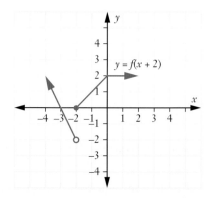

c   To go from             $y = f(x)$
    to                     $y = 2f(x)$
    involves multiplying the right hand side by 2.

    Thus the graph of      $y = 2f(x)$
    will be that of        $y = f(x)$
    dilated parallel to the $y$-axis, scale factor 2.

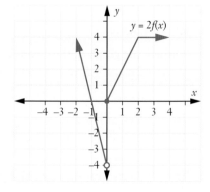

       ISBN 9780170390330

**d** To go from $y = f(x)$
to $y = f(0.5x)$
involves replacing $x$ by $0.5x$.

Thus the graph of $y = f(0.5x)$
will be that of $y = f(x)$

dilated parallel to the $x$-axis, scale factor $\dfrac{1}{0.5} = 2$.

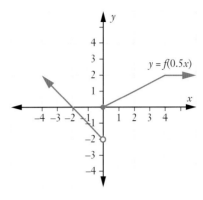

**e** To go from $y = f(x)$
to $y = f(2x)$
involves replacing $x$ by $2x$.

Thus the graph of $y = f(2x)$
will be that of $y = f(x)$

dilated parallel to the $x$-axis, scale factor $\dfrac{1}{2} = 0.5$.

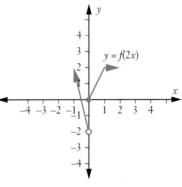

**f** To go from $y = f(x)$
to $y = -f(x)$
involves multiplying the right hand side by $-1$.

Thus the graph of $y = -f(x)$
will be that of $y = f(x)$
reflected in the $x$-axis.

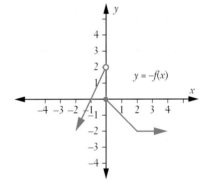

## Exercise 7C

**1** Describe how the graphs of each of the following can be obtained by transforming the graph of $y = f(x)$.

   **a** $y = -f(x)$       **b** $y = f(4x)$       **c** $y = 4f(x)$

**2** Describe how the graphs of each of the following can be obtained by transforming the graph of $y = x^2 + 3x$.

   **a** $y = -x^2 - 3x$       **b** $y = x^2 + 3x - 5$       **c** $y = \dfrac{x^2}{4} + \dfrac{3x}{2}$

**3** Describe how the graphs of each of the following can be obtained by transforming the graph of $y = x^2$.

**a** $y = (x - 3)^2$        **b** $y = 3x^2$        **c** $y = (3x)^2$

**4** The graph of $y = f(x)$ is shown on the right.
Draw the graph of each of the following.

**a** $y = f(x - 2)$
**b** $y = f(x) + 2$
**c** $y = 2f(x)$
**d** $y = f(2x)$
**e** $y = -f(x)$
**f** $y = f(-x)$

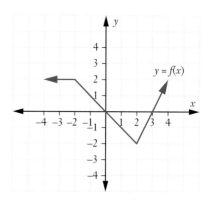

**5** The graph of $y = f(x)$ is shown on the right.
Find

**a** $f(0)$, i.e. the value of $y$ when $x = 0$.
**b** $f(1)$, i.e. the value of $y$ when $x = 1$.
**c** $f(2)$        **d** $f(-3)$

Draw the graph of each of the following.

**e** $y = f(x + 1)$        **f** $y = f(-x)$
**g** $y = f(2x)$        **h** $y = f(0.5x)$
**i** $y = 0.5f(x)$

**j** Use your part **b** answer and your part **e** graph to confirm that $f(1) = f(0 + 1)$.
**k** Use your part **c** answer and your part **g** graph to confirm that $f(2) = f(2 \times 1)$.

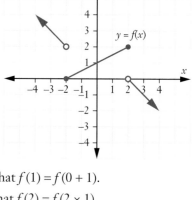

**6** The graph of $y = f(x)$ is as shown on the right.
Choose the function from the list below corresponding to each of the graphs A to F shown on the next page.

I     $y = -f(x)$
II    $y = f(-x)$
III   $y = 0.5f(x)$
IV   $y = f(0.5x)$
V     $y = 2f(x)$
VI    $y = f(2x)$
VII   $y = f(x) + 2$
VIII $y = f(x + 2)$
IX    $y = f(x) - 2$
X     $y = f(x - 2)$

**A**

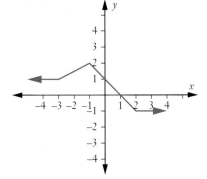

**B**

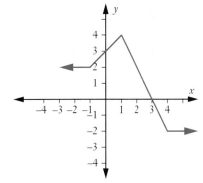

**C**

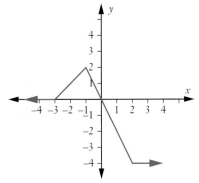

**D**

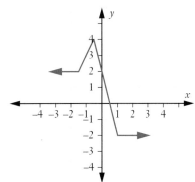

**E**

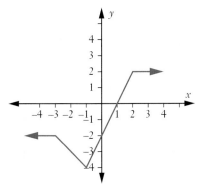

**F**

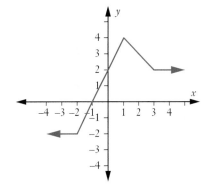

**7** The graph of $y = f(x)$ shown on the right, cuts the $x$-axis at A($-2$, 0), D(4, 0) and F(7, 0), cuts the $y$-axis at B(0, 4), has a maximum turning point at C(2, 5) and a minimum turning point at E(5, $-1$).

Find the coordinates of the points where

**a** $y = f(x - 3)$ cuts the $x$-axis,

**b** $y = f(2x)$ cuts the $x$-axis,

**c** $y = -f(x)$ cuts the $x$-axis,

**d** $y = f(-x)$ cuts the $x$-axis,

**e** $y = f(x) + 3$ has its maximum turning point,

**f** $y = -f(x)$ has its maximum turning point.

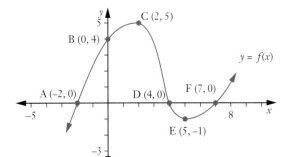

# Two relationships that are not functions

Consider the graph shown on the right.

The graph certainly looks parabolic in shape but is it a *quadratic function*?

Recalling from Chapter 3 the requirement that to be the graph of a function the graph must pass the *vertical line test*, the graph shown cannot be that of a function.

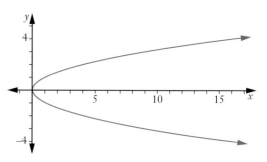

However the graph *is* parabolic in nature and its similarity with the graph of $y = x^2$ is because if we switch the $x$ and $y$ in $y = x^2$, to obtain $x = y^2$ (or $y^2 = x$), we have the equation of the given graph.

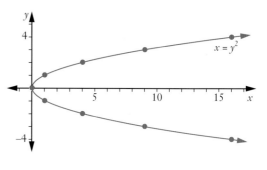

| $y$ | –4 | –3 | –2 | –1 | 0 | 1 | 2 | 3 | 4 |
|---|---|---|---|---|---|---|---|---|---|
| $x$ | 16 | 9 | 4 | 1 | 0 | 1 | 4 | 9 | 16 |

The graph of the relationship $y^2 = x$ is parabolic and has the $x$-axis as its line of symmetry.

Now consider the graphs shown below.

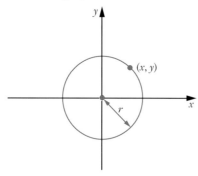

Circle centre (0, 0), radius $r$.

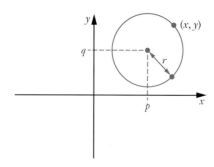

Circle centre ($p$, $q$), radius $r$.

Again each is *not* a function because for some $x$ values there exists more than one $y$ value. I.e. each relationship is 'one-to-many'. Each fails the vertical line test and therefore neither relationship is a function. However, as with $y^2 = x$ above, we can still determine a rule for each relationship.

Remember that the rule for a relationship is like the 'membership ticket' for the relationship – all points $(x, y)$ lying on the graph of the relationship will 'fit' the rule and all points not on the graph of the relationship will not fit the rule. For the above two relationships we can use the fact that in each case, all points lying on the circles must be a distance $r$ from the centre of the circle (and any point that is not a distance of $r$ from the centre will not lie on the circle) to determine a rule for each relationship. Thus, using the Pythagorean rule for right triangles, we have the rules:

$$x^2 + y^2 = r^2 \qquad \text{and} \qquad (x - p)^2 + (y - q)^2 = r^2$$
for a circle centre (0, 0) and radius $r$,      for a circle centre ($p$, $q$) and radius $r$.

If we expand

$$(x - p)^2 + (y - q)^2 = r^2$$

we obtain

$$x^2 - 2px + p^2 + y^2 - 2qy + q^2 = r^2$$

i.e.

$$x^2 + y^2 - 2px - 2qy = r^2 - p^2 - q^2$$

i.e.

$$x^2 + y^2 - 2px - 2qy = \text{(a constant)}$$

Notice that in this expanded form the Cartesian equation of a circle is characterised by:

- the coefficient of $x^2$ being the same as the coefficient of $y^2$,
- the only terms are those of $x^2, y^2, x, y$ and a constant (and of these any two of the last three could be zero).

## EXAMPLE 7

Find

**a** the equation of the circle centre $(3, -1)$ and radius 4.

**b** the centre and radius of the circle with equation: $x^2 + y^2 + 6y = 10x$.

### Solution

**a** The equation of a circle centre $(p, q)$ and radius $r$ is $(x - p)^2 + (y - q)^2 = r^2$

∴ The equation of a circle centre $(3, -1)$ and radius 4 is $(x - 3)^2 + (y + 1)^2 = 16$

**b** We need to rearrange the given equation to the form $(x - p)^2 + (y - q)^2 = r^2$

$$x^2 + y^2 + 6y = 10x$$
$$x^2 - 10x + \ldots + y^2 + 6y + \ldots = 0 \qquad \leftarrow \text{create gaps}$$
$$x^2 - 10x + 25 + y^2 + 6y + 9 = 0 + 25 + 9 \quad \leftarrow \text{complete the squares}$$
$$(x - 5)^2 + (y + 3)^2 = 34$$

Comparing with $(x - p)^2 + (y - q)^2 = r^2$ we see that the given circle has its centre at $(5, -3)$ and a radius of $\sqrt{34}$.

## Exercise 7D

**1** Which of the following equations represent circles?

A: $x^2 + y^2 - 2x + 4y = 6$     B: $2x^2 + y^2 - 3x + 8y + 10 = 0$     C: $x^2 + y^2 = 6$

D: $x^2 + y^2 + 8x = 10$     E: $x^2 - y^2 + 2x + 10y = 100$     F: $x^2 + 6xy + y^2 + 15y = 20$

**2** Find the equation of a circle centre $(0, 0)$ and radius 10 units.

If each of the following points lie on this circle, determine $a, b, c$ and $d$ given that $a$ and $b$ are positive and $c$ and $d$ are negative.

Point A$(-6, a)$,     Point B$(3, b)$,     Point C$(0, c)$,     Point D$(d, 5)$.

**3** Find the equation of each of the following circles, giving your answers in the form

$$(x - p)^2 + (y - q)^2 = c.$$

**a** Centre $(2, -3)$ and radius 5.     **b** Centre $(3, 2)$ and radius 7.

**c** Centre $(-10, 2)$ and radius $3\sqrt{5}$.     **d** Centre $(-1, -1)$ and radius 6.

**4** Find the equation of each of the following circles, giving your answers in the form
$$x^2 + y^2 + dx + ey = c.$$

**a** Centre $(3, 5)$ and radius 5.      **b** Centre $(-2, 1)$ and radius $\sqrt{7}$.

**c** Centre $(-3, -1)$ and radius 2.      **d** Centre $(3, 8)$ and radius $2\sqrt{7}$.

**5** Find the radius and the coordinates of the centre of each of the following circles.

**a** $x^2 + y^2 = 25$      **b** $25x^2 + 25y^2 = 9$

**c** $(x - 3)^2 + (y + 4)^2 = 25$      **d** $(x + 7)^2 + (y - 1)^2 = 100$

**e** $x^2 + y^2 - 6x + 4y + 4 = 0$      **f** $x^2 + y^2 + 2x - 6y = 15$

**g** $x^2 + y^2 + 2x = 14y + 50$      **h** $x^2 + 10x + y^2 = 151 + 14y$

**i** $x^2 + y^2 = 20x + 10y + 19$      **j** $2x^2 - 2x + 2y^2 + 10y = -5$

**6** Find the distance between the centres of the two circles given below:
$$(x - 3)^2 + (y - 7)^2 = 36 \quad \text{and} \quad (x - 2)^2 + (y - 9)^2 = 49.$$

**7** The circle $(x - 3)^2 + (y + 4)^2 = 25$ has its centre at point A.

The circle $(x - 2)^2 + (y - 7)^2 = 9$ has its centre at point B.

Find the equation of the straight line through A and B.

**8** The circle $(x + 1)^2 + (y - 7)^2 = 36$ is moved right 4 units and down 3 units. What will be the equation of the circle in its new location?

**9** The circle $x^2 + y^2 - 6x + 10y + 25 = 0$ is moved left 7 units and up 2 units. What will be the equation of the circle in its new location?

**10** Each of the following graphs show $y^2 = x$ translated up, down, left or right.

Determine the equation of each relationship shown.

**a** 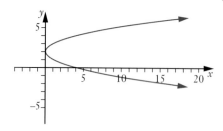      **b**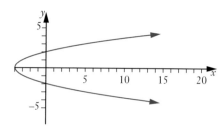

**c**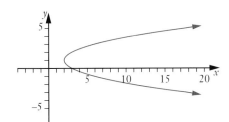

**d**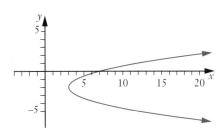

**11** Point A is the centre of the circle $(x - 3)^2 + (y - 11)^2 = 144$.

Point B is the centre of the circle $(x - 12)^2 + (y + 1)^2 = 9$.

  **a**  Determine the length of AB.

  **b**  Determine whether the circles have two points in common, just one point in common or no points in common, and justify your answer.

**12** Point C is the centre of the circle $(x - 2)^2 + (y - 3)^2 = 9$.

Point D is the centre of the circle $(x + 2)^2 + (y - 5)^2 = 1$.

  **a**  Determine the length of CD.

  **b**  Determine whether the circles have two points in common, just one point in common or no points in common, and justify your answer.

**13** Find the coordinates of the points where the line $y = x - 3$ meets the circle
$$(x - 4)^2 + (y - 2)^2 = 25.$$

**14** Find the coordinates of the points where the line $4y = x + 30$ meets the circle
$$(x + 5)^2 + (y - 2)^2 = 34.$$

**15** Prove that the straight line $3y = x + 25$ is a tangent to the circle
$$(x - 7)^2 + (y - 4)^2 = 40,$$

and find the coordinates of the point of contact.

**16** What restriction is there on the possible values of $a$ if
$$x^2 + 2x + y^2 - 10y + a = 0$$

is the equation of a circle?

**7.** Polynomials and other functions ●●●●●●●

# Miscellaneous exercise seven

This miscellaneous exercise may include questions involving the work of this chapter, the work of any previous chapters, and the ideas mentioned in the Preliminary work section at the beginning of the book.

**1** Determine the equation of each of the following cubic functions.

**a**

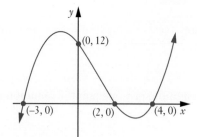

**b**

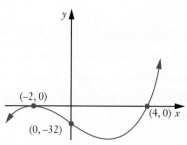

**2** Without the assistance of a calculator, solve the following quadratic equation
$$x^2 - 4x - 6 = 0$$

**a** using the quadratic formula

**b** by completing the square, and in each case give answers in exact form.

**3** Use the technique of completing the square to determine the centre and radius of the circle with equation $x^2 + 6x + y^2 - 10y = 15$.

**4** If $f(x) = x$, $g(x) = x^2$ and $h(x) = x^3$ determine

**a** $f(4)$        **b** $g(4)$        **c** $h(4)$

**d** the values of $p$ for which $f(p)$, $g(p)$ and $h(p)$ are all the same

**5** State which of the following functions are linear and, for those that are, state the gradient.

$f_1(x): y = \sqrt{x}$                  $f_2(x): 2y = 5x + 4$

$f_3(x): y = (x + 1)(x + 3)$          $f_4(x): 2x + y = 3$

**6** Find the equation of the straight line that passes through the point $(15, -1)$ and that is perpendicular to the line $5x + 2y = 9$.

**7** Use either the graphing facility or the equation solving facility of a calculator to solve each of the following equations for $x \in \mathbb{R}$.

**a** $8x^3 + 18x^2 - 221x + 315 = 0$

**b** $8x^3 - 2x^2 = 441 + 315x$

**c** $2x^3 - 11x^2 + 19x = 12$

**d** $x^4 - 3x^3 + 12x^2 - 21x + 35 = 0$

**8** For each of parts **a** to **h** below, state which of the following statements apply:

Statement A:   $y$ increases as $x$ increases.

Statement B:   $y$ decreases as $x$ increases.

Statement C:   $y$ is directly proportional to $x$.

Statement D:   $y$ is inversely proportional to $x$.

**a**   $y = 5x$      **b**   $y = \dfrac{7}{x}$      **c**   $y = \dfrac{2}{x}$      **d**   $y = \dfrac{x}{3}$      **e**   $y = 2x + 1$

**f** 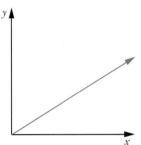    **g** 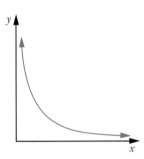    **h**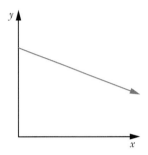

**9** Without the help of a calculator, solve each of the following equations for $x \in \mathbb{R}$.

**a**   $(2x - 7)(x + 9) = 0$      **b**   $x^2 - 8x + 12 = 0$      **c**   $5x^2 + 2x - 3 = 0$

**d**   $(x + 11)(5x - 4)(x - 7) = 0$      **e**   $(x - 3)(x^2 + 4x - 5) = 0$      **f**   $(x + 5)(2x^2 + x - 6) = 0$

**10** Classify each of the following as appearing to be the graph of a linear function, or a quadratic function, or a cubic function, or a reciprocal function or none of the previous.

**a** 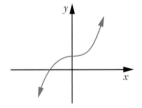    **b** 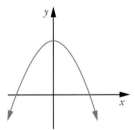    **c**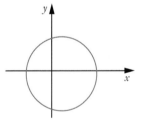

**d** 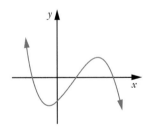    **e** 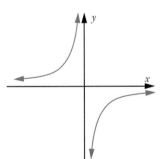    **f**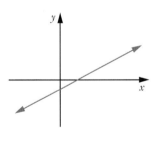

**11** Given that $x^3 - 8x^2 + 19x - 12 = (x - 3)(x^2 + bx + c)$:

   **a**   determine the value of $c$ by inspection

   **b**   with your answer from part **a** in place of $c$, expand $(x - 3)(x^2 + bx + c)$ and hence determine $b$

   **c**   factorise $x^3 - 8x^2 + 19x - 12$

**12**

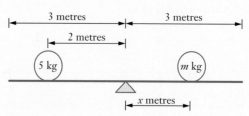

To balance the system above the relationship between $m$ and $x$ must be:

$$m = \frac{10}{x}.$$

   **a**   If $x$ is doubled in value, what must happen to the value of $m$ if the system is to remain in balance?

   **b**   State whether the relationship between $x$ and $m$ is one of direct proportion, inverse proportion or neither of these and explain what your answer means in terms of the way $m$ needs to vary as $x$ is varied if the system is to remain in balance.

   **c**   If $m = 20$ what must be the value of $x$ for the system to be in balance?

   **d**   With the $x$ values as input, the $m$ values as output and the system in balance, what is the domain and range for this function?

**13** An engineering component consists of a rectangular metal plate with a triangular piece removed, as in the diagram below left. The removed piece is cut away by a computer controlled machine that is programmed to cut a triangle with vertices at the distances and angles shown on the diagram below right.

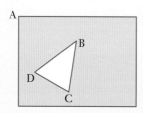

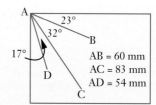

Find the area and the perimeter of the triangular piece that is removed.

ISBN 9780170390330

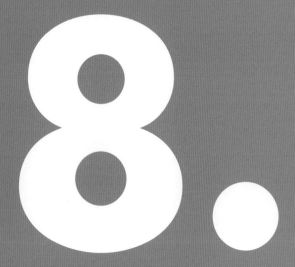

# Trigonometric functions

# The graph of y = sin x

In the diagram below, a circle of unit radius has been divided into sectors each of central angle 15°. The graph has been obtained by plotting on the $y$ axis the height of point A above the centre of the circle as A moves around the unit circle in an anticlockwise sense, against the angle moved through on the $x$ axis. The dots on the graph are plotted every 15°. From the unit circle definition of the sine of an angle, encountered in Chapter 1, it follows that the graph produced in this way is that of $y = \sin x$.

Sine and cosine curves

Sketching periodic functions-amplitude and period

Sketching periodic functions-phase and vertical shift

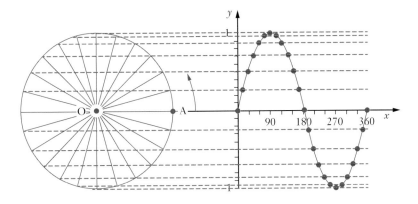

If, having completed one rotation of the circle, we were to continue moving point A around the circle the graph would repeat itself, as shown below for three rotations.

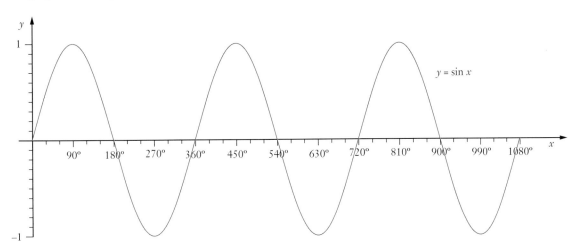

iStock.com/Wojciech Kozielczyk

Alternatively angles could be shown in radians and negative angles could also be included, as shown below for $-2\pi \le x \le 4\pi$.

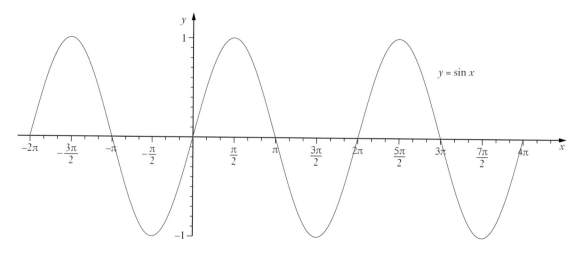

Whilst the graph shown above is for $-2\pi \le x \le 4\pi$, this restriction is made purely due to page width limitations. The reader should consider the graph of $y = \sin x$ continuing indefinitely to the left and the right.

Points to note:

- The graph of $y = \sin x$ repeats itself every $2\pi$ radians (or 360°).
  We say that the sine function is **periodic**, with **period** $2\pi$.
  Thus $\sin(x \pm 2\pi) = \sin x$.

- We also say that the graph performs one **cycle** each period.
  Thus $y = \sin x$ performs one cycle in $2\pi$ radians (or 360°).

- Note that $-1 \le \sin x \le 1$.

- If we consider the above graph to have a 'mean' $y$-coordinate of $y = 0$ then the graph has a maximum value 1 above this mean value and a minimum value 1 below it. We say that $y = \sin x$ has an **amplitude** of 1.

- The graph passes the 'vertical line test', i.e. for each $x$ value there is one and only one $y$ value. Hence $y = \sin x$ is a function.

- Note that $\sin(-x) = -\sin x$. (Functions for which $f(-x) = -f(x)$ are called *odd* functions and are unchanged under a 180° rotation about the origin. As is the case for $y = x^n$ for *odd* values of $n$.)

Display the graph of $y = \sin x$ on your calculator and confirm that it is as shown above.

ISBN 9780170390330

# The graph of y = cos x

Similar considerations of the x-coordinate of point A as it moves around the unit circle gives the graph of

$$y = \cos x,$$

shown below for $-2\pi \le x \le 4\pi$.

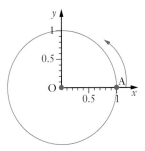

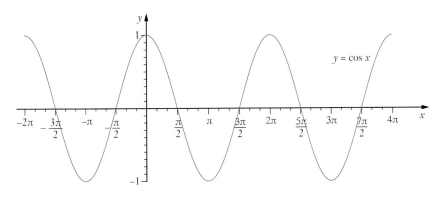

Points to note:

* The cosine function is **periodic**, with **period 2π**. Thus cos (x ± 2π) = cos x.

* y = cos x performs one **cycle** in 2π radians (or 360°).

* Note that −1 ≤ cos x ≤ 1.

* The graph of y = cos x has an **amplitude** of 1.

* Note that cos (−x) = cos x. (Functions for which $f(-x) = f(x)$ are called *even* functions and are unchanged under a reflection in the y-axis. As is the case for $y = x^n$ for *even* values of n.)

* If the above graph of y = cos x is moved $\dfrac{\pi}{2}$ units right, parallel to the x-axis, it would then be the same as the graph of y = sin x. We say that sin x and cos x are $\dfrac{\pi}{2}$ out of **phase** with each other.

  It follows that $\cos x = \sin\left(x + \dfrac{\pi}{2}\right)$ and $\sin x = \cos\left(x - \dfrac{\pi}{2}\right)$.

---

**Note**

Whether an integer is even or odd is called the **parity** of the integer. Thus two odd numbers have the same parity. Similarly, whether a function is even, i.e. $f(-x) = f(x)$, or odd, i.e. $f(-x) = -f(x)$, is referred to as the parity of the function.

A function does not have to be even or odd. Many functions are neither even nor odd, (and indeed the function $f(x) = 0$ can be regarded as being *both* even and odd).

---

# The graph of y = tan x

The fact that $\tan x = \dfrac{\sin x}{\cos x}$ was justified for right triangles in the *Preliminary work*.

If we assume this relationship to be true for all angles then it follows that $y = \tan x$ will

- equal zero for any value of $x$ for which $\sin x = 0$, i.e. $x = 0°, \pm180°, \pm360°, \ldots$

- be undefined for any value of $x$ for which $\cos x = 0$, i.e. $x = \pm90°, \pm270°, \ldots$

- equal 1 for any value of $x$ for which $\sin x = \cos x$, eg. $x = 45°, 225°, 405°, \ldots$

- equal $-1$ for any value of $x$ for which $\sin x = -\cos x$, eg. $x = 135°, 315°, \ldots$ .

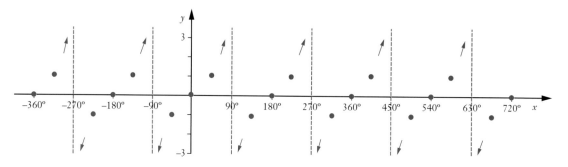

The completed graph of $y = \tan x$, for $-2\pi \le x \le 4\pi$, is shown below.

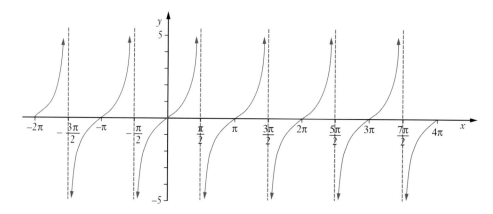

Note:

- Though the graph above is for $-2\pi \le x \le 4\pi$ the reader should consider the graph of $y = \tan x$ continuing indefinitely to the left and right.

- The graph repeats itself every $\pi$ radians (or 180°).
  The **period** of the graph is $\pi$ radians (or 180°). Thus $\tan (x \pm \pi) = \tan x$.
  The graph performs one **cycle** in $\pi$ radians (or 180°).

- The term 'amplitude' is meaningless when applied to $y = \tan x$.

- The graph is such that $\tan (-x) = -\tan x$. (The tangent function is an *odd* function.)

ISBN 9780170390330

# More about $y = \tan x$

The previous page developed the graph of $y = \tan x$ from the rule
$\tan x = \dfrac{\sin x}{\cos x}$.

Alternatively we can use the unit circle to define the tangent of an
angle directly. Consider some point A on the unit circle, centre at
point O(0, 0). Point B is the point (1, 0). We will define the tangent of
the angle AOB (shown as a 50° angle in the diagram) as the $y$-coordinate
of the point where OA, continued as necessary, meets the vertical line
through B. (This is point C in the diagram on the right and gives tan
∠AOB as just less than 1.2, i.e. tan 50° ≈ 1.2.)

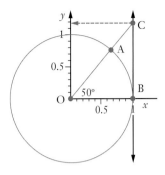

Note:

- The vertical through B is the tangent to the circle at point B so this definition of the trigonometric term *tangent* does involve a *tangent* to the unit circle.

- Applying the right triangle definition of tangent to triangle OBC, we also obtain tan ∠AOB = BC, so the two approaches are consistent.

The diagram below shows this definition used to produce the graph of $y = \tan x$ for $x$ from 0° to 180°.
The reader should confirm that this graph:

**a**  is consistent with the graph produced previously using the unit circle definitions of sin $x$ and cos $x$
and the relationship $\tan x = \dfrac{\sin x}{\cos x}$

**b**  would repeat every 180° thereafter.

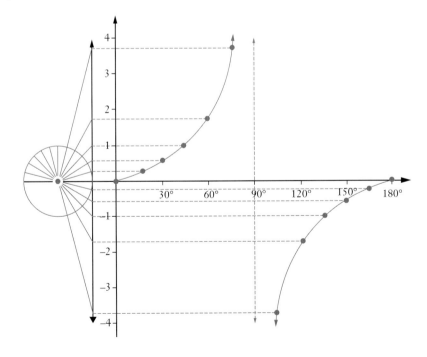

## EXAMPLE 1

State the amplitude and period of the function shown graphed on the right.

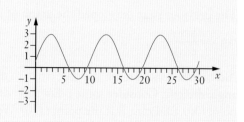

**Solution**

The graph appears to have a maximum value of 3, a minimum value of –1. These values are respectively 2 units above and 2 units below the 'mean' value of 1. The amplitude is 2.

The graph appears to repeat itself every 10 units.

The period is 10.

## Exercise 8A

State the period of each of the periodic functions shown graphed below.

**1**
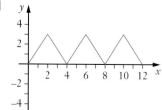

**2**
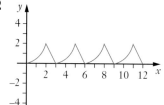

**3**
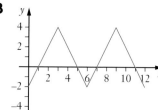

**4**
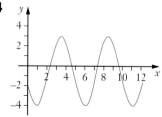

**5**
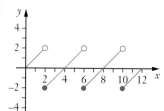

**6**
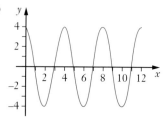

Alamy Stock Photo/Kenna Love

State the amplitude and period of each of the functions shown below.

**7**
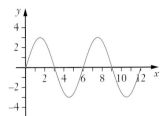

**8**
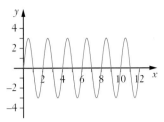

**9**
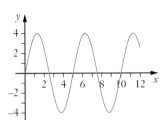

**10**
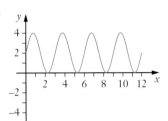

**11**
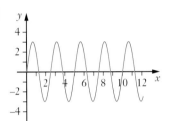

**12**
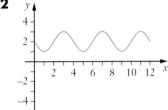

## INVESTIGATION

- What effect does changing the value of $a$ in $y = a \sin x$ have on the graph of the function?
  I.e. draw and compare $y = \sin x$, $y = 2 \sin x$, $y = 3 \sin x$, $y = -2 \sin x$ etc. and try to write a general statement regarding the effect altering $a$ has on the graph of $y = a \sin x$.

- Similarly investigate changing the value of $b$ in $y = a \sin (bx)$, for $b > 0$.
  I.e. for some fixed value of $a$, say 1, compare graphs of $y = 1 \sin x$, $y = 1 \sin 2x$, $y = 1 \sin 3x$, $y = 1 \sin 4x$ etc. and try to write a general statement regarding the effect altering $b$ has on the graph of $y = a \sin bx$.

- Similarly investigate changing the value of $c$ in $y = a \sin [b(x - c)]$
  and $d$ in $y = a \sin [b(x - c)] + d$.

- Similarly investigate the cosine and tangent functions.

EXAMPLE 2

With $x$ in degrees the graph on the right shows
$$y = \cos x$$
and $\quad y = a \cos x.$

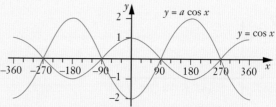

The second graph, below right, shows
$$y = a \cos (x - b)$$
for $b$ in degrees and
$$0 \le b \le 360.$$

Determine $a$ and $b$.

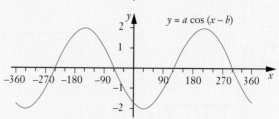

### Solution

In the first diagram, the graph of $y = a \cos x$ is that of $y = \cos x$ stretched $\updownarrow$ parallel to the $y$-axis until its amplitude is 2, and then reflected in the $x$-axis. This effect will be achieved if $a = -2$. The second diagram shows the graph of $y = -2 \cos x$ moved $30°$ right. This is achieved if $b = 30$. Hence $a = -2$ and $b = 30$.

EXAMPLE 3

With $x$ in radians, the graph below left shows $y = \sin bx$ and $y = a \sin bx$.

The graph below right shows $y = a \sin [b(x - c)]$, $0 \le c \le \pi$. Find $a$, $b$ and $c$.

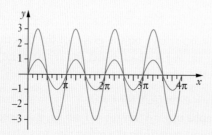 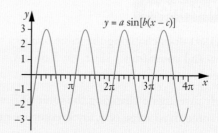

### Solution

The 'smaller' line in the first graph, with an amplitude of 1, must be $y = \sin bx$.

The line performs 2 cycles in the interval that $y = \sin x$ would perform 1. Hence $b = 2$.

(Alternatively we could say: $y = \sin bx$ has a period of $\dfrac{2\pi}{b}$. The given graph has a period of $\pi$.

$\qquad$ Thus $\dfrac{2\pi}{b} = \pi$, which again gives $b = 2$.)

The other line in the first graph, is $y = \sin 2x$ stretched $\updownarrow$ parallel to the $y$-axis until its amplitude is 3. Hence $a = 3$.

The second graph shows $y = 3 \sin 2x$ moved $\dfrac{\pi}{8}$ units right. Hence $a = 3$, $b = 2$ and $c = \dfrac{\pi}{8}$.

ISBN 9780170390330

EXAMPLE 4

Sketch the graph of $y = 2 \tan \dfrac{x}{2}$ for $0 \le x \le 4\pi$ and then check the reasonableness of your sketch by viewing the graph of the function on your calculator.

**Solution**

The graph of $y = \tan \dfrac{x}{2}$ will perform half of a cycle in the interval that $y = \tan x$ would perform 1 cycle, i.e. $\pi$ radians. Thus $\tan \dfrac{x}{2}$ has a period of $2\pi$ radians.

Also (remembering that $\tan \dfrac{\pi}{4} = 1$) when $x = \dfrac{\pi}{2}$, $y = 1$.

The graph of $y = \tan \dfrac{x}{2}$ is as shown by the red broken lines in the diagram on the right.

Stretching $y = \tan \dfrac{x}{2}$ parallel to the $y$-axis ($\updownarrow$), until distances from the $y$-axis are doubled, will give the graph of $y = 2 \tan \dfrac{x}{2}$, as shown by the blue solid lines in the graph. (The reader should check the reasonableness of the sketch by comparing it to that from a calculator display.)

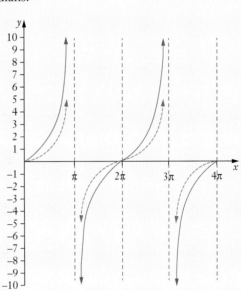

## Exercise 8B

Attempt the following without the assistance of a graphic calculator, then use your calculator to check your answers if you wish.

**1** State the amplitude of each of the following.

**a** $y = \sin x$      **b** $y = 2 \cos x$      **c** $y = 4 \cos x$

**d** $y = -3 \sin 2x$      **e** $y = 2 \cos \left( x + \dfrac{\pi}{2} \right)$      **f** $y = -3 \sin (x - \pi)$

**g** $y = 5 \cos (x - 2)$      **h** $y = -3 \cos (2x + \pi)$

**2** State the period of each of the following for $x$ in degrees.

**a** $y = \sin x$      **b** $y = \tan x$      **c** $y = 2 \sin x$

**d** $y = \sin 2x$      **e** $y = \cos \dfrac{x}{2}$      **f** $y = \cos 3x$

**g** $y = 3 \tan 2x$      **h** $y = 3 \sin \left( \dfrac{x - 60°}{3} \right)$      **i** $y = 5 \sin [2(x - 30°)]$

**3** State the period of each of the following for $x$ in radians.

   **a**   $y = \cos x$             **b**   $y = \tan x$             **c**   $y = 3 \cos x$

   **d**   $y = 2 \cos 4x$          **e**   $y = 2 \tan 3x$         **f**   $y = \dfrac{1}{2} \sin 3x$

   **g**   $y = 3 \sin\left(\dfrac{x}{2}\right)$      **h**   $y = 2 \cos (2x - \pi)$      **i**   $y = 2 \sin (4\pi x)$

**4** Determine the coordinates of any maximum and minimum points on each of the following functions for $0 \le x \le 2\pi$.

   **a**   $y = \sin x$             **b**   $y = 2 + \sin x$         **c**   $y = -\sin x$

   **d**   $y = \sin (2x) + 3$       **e**   $y = \sin\left(x - \dfrac{\pi}{4}\right) + 3$

**5** State the greatest value each of the following can take and the smallest positive value of $x$ (in degrees) that gives this maximum value.

   **a**   $3 \sin x$       **b**   $2 \sin (x - 30°)$      **c**   $2 \sin (x + 30°)$      **d**   $-3 \sin x$

**6** State the greatest value each of the following can take and the smallest positive value of $x$ (in radians) that gives this maximum value.

   **a**   $3 \sin 2x$       **b**   $-5 \sin x$       **c**   $2 \cos\left(x + \dfrac{\pi}{6}\right)$      **d**   $3 \cos\left(x - \dfrac{\pi}{6}\right)$

**7** Each of the following graphs has an equation of the form $y = a \sin x$. State the value of $a$ in each case.

   **a**
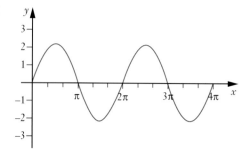

   **b**
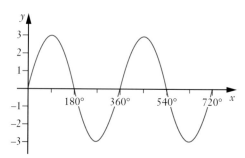

   **c**
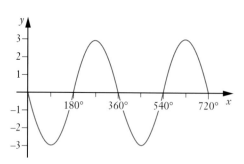

   **d**
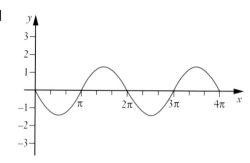

ISBN 9780170390330

**8** Each of the following graphs has an equation of the form $y = a \cos x$.

State the value of $a$ in each case.

**a**

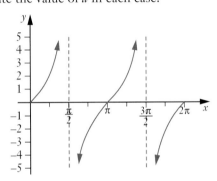

**b**

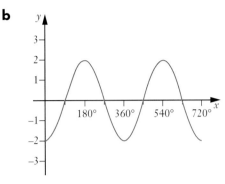

**9** Each of the following graphs has an equation of the form $y = a \tan x$.

State the value of $a$ in each case.

**a**

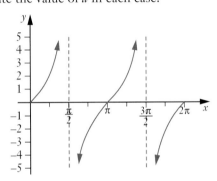

**b**

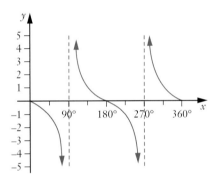

**10** Each of the following graphs has an equation of the form $y = a \sin bx$.

State the values of $a$ and $b$ in each case.

**a**

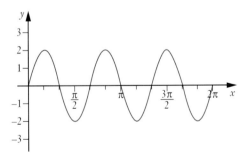

**b**

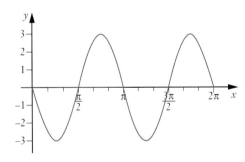

**c**

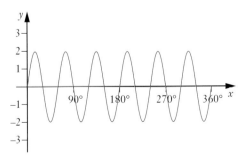

**d**

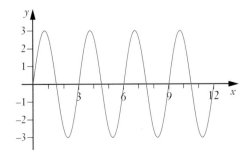

**8.** Trigonometric functions ●●●●●●●●

**11** Each of the following graphs has an equation of the form $y = a \cos bx$.
State the value of $a$ and $b$ in each case.

**a**
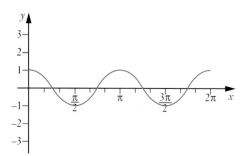

**b**
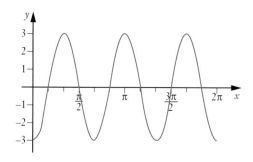

**c**
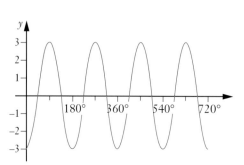

**d**
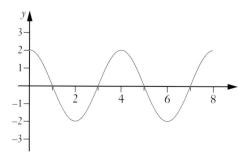

**12** The graph on the right shows $y = a \sin x°$ and $y = a \sin (x - b)°$, with $a$ the same integer in both equations, $b$ a multiple of ten and one of the lines shown as a red line.

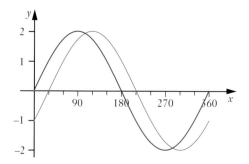

    **a**    Find the value of $a$ and the two smallest possible positive values of $b$.

    **b**    Give an equation for the blue line in the form $y = c \sin (x - d)°$ with $c$ a negative integer, $d$ a multiple of ten and $0 \leq d \leq 360$.

**13**  **a**    State the period and amplitude of $y = 3 \cos (\pi x)$.

    **b**    Sketch the graph of $y = 3 \cos (\pi x)$ for $0 \leq x \leq 6$.

**14**  **a**    State the period and amplitude of $y = -5 \sin \left( \dfrac{\pi x}{2} \right)$.

    **b**    Sketch the graph of $y = -5 \sin \left( \dfrac{\pi x}{2} \right)$ for $0 \leq x \leq 8$.

**15** Sketch both of the following on a single pair of axes, with $0 \leq x \leq 360°$.

    **a**    $y = 2 \tan x$                         **b**    $y = 2 \tan (x + 45°)$

**16** Sketch both of the following on a single pair of axes, with $0 \leq x \leq 2\pi$.

    **a**    $y = 3 \sin 2x$                  **b**    $y = 3 \sin \left( 2x - \dfrac{\pi}{3} \right)$, i.e. $y = 3 \sin \left[ 2\left( x - \dfrac{\pi}{6} \right) \right]$

ISBN 9780170390330

# Positive or negative?

As we saw in Chapter 7, the *x*- and *y*-axes divide the coordinate plane into four regions called **quadrants**. These four quadrants are numbered as shown on the right.

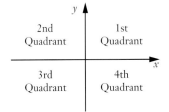

- Any points lying in the 1st or 4th quadrants will have a positive *x*-coordinate.

- Any points lying in the 2nd or 3rd quadrants will have a negative *x*-coordinate.

With our unit circle definition for the cosine of an angle involving the *x*-coordinate of a point moving around the circle, it follows that angles with their *initial ray* along the positive *x*-axis, and their *terminal ray* lying in either the 1st or 4th quadrants, will have positive cosines and any with their terminal ray lying in the 2nd or 3rd quadrants will have cosines that are negative.

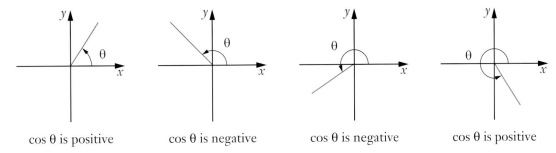

| cos θ is positive | cos θ is negative | cos θ is negative | cos θ is positive |

This positive or negative nature of the cosine function can be summarised as shown in the diagram on the right.

| Cosine negative | Cosine positive |
|---|---|
| Cosine negative | Cosine positive |

Similarly, remembering that the unit circle definition of the sine of an angle involves the *y*-coordinate of a point moving around the circle, the situation for sine is as follows:

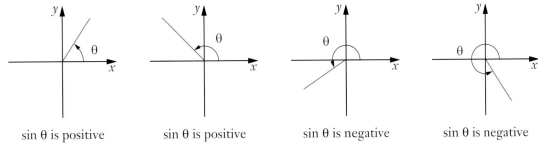

| sin θ is positive | sin θ is positive | sin θ is negative | sin θ is negative |

This positive or negative nature of the sine function can be summarised in the diagram on the right.

| Sine positive | Sine positive |
|---|---|
| Sine negative | Sine negative |

From $\tan \theta = \dfrac{\sin \theta}{\cos \theta}$ it follows from these facts about sine and cosine that the positive and negative nature of tan θ will be as shown in the diagram on the right.

| Tangent negative | Tangent positive |
|---|---|
| Tangent positive | Tangent negative |

These facts regarding the positive and negative nature of sine, cosine and tangent are summarised below.

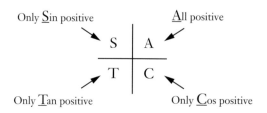

Only <u>S</u>in positive     <u>A</u>ll positive

S | A

T | C

Only <u>T</u>an positive     Only <u>C</u>os positive

## EXAMPLE 5

Without the assistance of a calculator state whether each of the following are positive or negative.

**a**    sin 240°       **b**    cos 170°       **c**    tan (1.2π)       **d**    $\sin\left(-\dfrac{\pi}{3}\right)$

### Solution

**a**

| Sin +ve | All +ve |
|---|---|
| Tan +ve | Cos +ve |

**b**

| Sin +ve | All +ve |
|---|---|
| Tan +ve | Cos +ve |

**c**

| Sin +ve | All +ve |
|---|---|
| Tan +ve | Cos +ve |

**d**

| Sin +ve | All +ve |
|---|---|
| Tan +ve | Cos +ve |

sin 240° is negative    cos 170° is negative    tan (1.2π) is positive    $\sin\left(-\dfrac{\pi}{3}\right)$ is negative

Consider angles α, β and θ as shown below.

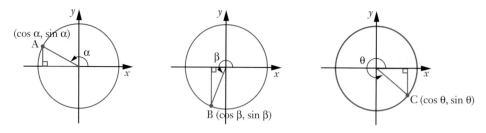

In each case, the right-angled triangle made with the *x*-axis could be re-drawn in the first quadrant as shown on the next page.

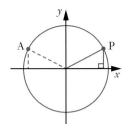 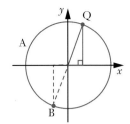 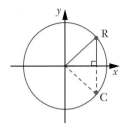

The $x$- and $y$-coordinates of P, Q and R may differ from those of A, B and C only in sign.

e.g. $x$-coordinate of P = $-x$-coordinate of A,
$y$-coordinate of P = $y$-coordinate of A.

Thus the sine (or cosine or tangent) of any angle will equal the sine (or cosine or tangent) of the acute angle made with the $x$-axis together with the appropriate sign.

## EXAMPLE 6

Express sin 200° in terms of the sine of an acute angle.

**Solution**

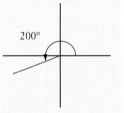

An angle of 200° makes 20° with the $x$-axis and lies in the 3rd quadrant, where the sine function is negative.

Thus sin 200° = $-$sin 20°.

## EXAMPLE 7

Express tan (−155°) in terms of the tangent of an acute angle.

**Solution**

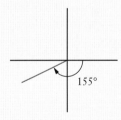

An angle of −155° makes 25° with the $x$-axis and lies in the 3rd quadrant, where the tangent function is positive.

Thus tan (−155°) = tan 25°.

## EXAMPLE 8

Find the exact value of cos 300°.

**Solution**

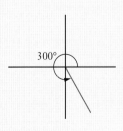

An angle of 300° makes 60° with the $x$-axis and lies in the 4th quadrant, where the cosine function is positive. Thus cos 300° = cos 60°

$$= \frac{1}{2}.$$

## EXAMPLE 9

Find the exact value of sin 270°.

**Solution**

An angle of 270° makes 90° with the x-axis and lies on the boundary between the 3rd and 4th quadrant. In both of these quadrants the sine function is negative.

Thus sin 270° = –sin 90°

   = –1.

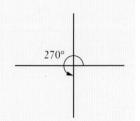

## EXAMPLE 10

Give the exact value of $\tan \dfrac{11\pi}{6}$.

**Solution**

An angle of $\dfrac{11\pi}{6}$ makes $\dfrac{\pi}{6}$ with the x-axis and lies in the 4th quadrant, where the tangent function is negative.

$$\tan \frac{11\pi}{6} = -\tan \frac{\pi}{6}$$

$$= -\frac{1}{\sqrt{3}} \qquad \text{(Or, expressed with a rational denominator, } -\frac{\sqrt{3}}{3}\text{.)}$$

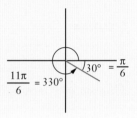

## Exercise 8C

**(Without the assistance of a calculator.)**

For each of the following state whether positive or negative.

**1**  tan 190°  **2**  cos 310°  **3**  tan (–190°)  **4**  sin (–170°)

**5**  sin 555°  **6**  cos 190°  **7**  $\tan \dfrac{\pi}{10}$  **8**  $\sin \dfrac{4\pi}{5}$

**9**  $\cos \dfrac{\pi}{10}$  **10**  $\sin\left(-\dfrac{\pi}{5}\right)$  **11**  $\cos \dfrac{9\pi}{10}$  **12**  $\tan \dfrac{13\pi}{5}$

Express each of the following in terms of the sine of an acute angle.

**13**  sin 140°  **14**  sin 250°  **15**  sin 340°  **16**  sin 460°

**17**  $\sin \dfrac{5\pi}{6}$  **18**  $\sin \dfrac{7\pi}{6}$  **19**  $\sin \dfrac{11\pi}{5}$  **20**  $\sin\left(-\dfrac{\pi}{5}\right)$

Express each of the following in terms of the cosine of an acute angle.

**21**  cos 100°  **22**  cos 200°  **23**  cos 300°  **24**  cos (–300°)

**25**  $\cos \dfrac{4\pi}{5}$  **26**  $\cos \dfrac{9\pi}{10}$  **27**  $\cos \dfrac{11\pi}{10}$  **28**  $\cos \dfrac{21\pi}{10}$

Express each of the following in terms of the tangent of an acute angle.

**29** $\tan 100°$        **30** $\tan 200°$        **31** $\tan(-60°)$        **32** $\tan(-160°)$

**33** $\tan \dfrac{6\pi}{5}$        **34** $\tan\left(-\dfrac{6\pi}{5}\right)$        **35** $\tan \dfrac{11\pi}{5}$        **36** $\tan\left(-\dfrac{21\pi}{5}\right)$

Give the exact value of each of the following.

**37** $\sin 300°$        **38** $\tan 210°$        **39** $\cos 240°$        **40** $\cos 270°$

**41** $\sin 180°$        **42** $\cos 390°$        **43** $\sin(-135°)$        **44** $\cos(-135°)$

**45** $\sin \dfrac{7\pi}{6}$        **46** $\cos \dfrac{7\pi}{6}$        **47** $\tan \dfrac{7\pi}{6}$        **48** $\sin \dfrac{7\pi}{4}$

**49** $\cos\left(-\dfrac{7\pi}{4}\right)$        **50** $\tan(6\pi)$        **51** $\sin \dfrac{5\pi}{2}$        **52** $\cos\left(-\dfrac{7\pi}{3}\right)$

# Solving trigonometric equations

Suppose we are asked to find an angle, $x$, such that $\sin x = 0.5$. With $\sin x$ being positive we know that any solutions must lie in the 1st and 2nd quadrants. The acute angle made with the $x$-axis must be 30° because, from our exact values, we know that $\sin 30° = 0.5$.

Thus, diagrammatically, the two possibilities for $x$ are as shown on the right.

However, if there is no restriction on $x$, there are an infinite number of values of $x$ that we can obtain from this diagram, and for all of these $\sin x = 0.5$. Twelve such values of $x$, six positive and six negative, are shown below. The reader should use a calculator to confirm that each of these values satisfy the requirement that $\sin x = 0.5$.

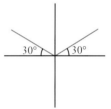

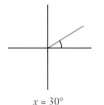
$x = 30°$

$x = 150°$

$x = 390°$

$x = 510°$

$x = 750°$

$x = 870°$

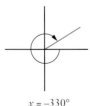
$x = -210°$          $x = -330°$          $x = -570°$

$x = -690°$          $x = -930°$

$x = -1050°$

Thus when asked to solve trigonometrical equations we will usually be given certain restrictions on the range of values the solutions can take.

Without the assistance of a calculator, solve $\sin x = -\dfrac{\sqrt{3}}{2}$ for $0 \le x \le 360°$.

**Solution**

With the sine being negative, solutions must lie in the 3rd and 4th quadrants.

From our exact values we know that $\sin 60° = \dfrac{\sqrt{3}}{2}$.

Thus the solutions make $60°$ with the $x$-axis as shown diagrammatically on the right.

Using this diagram to obtain solutions in the required interval we have $x = 240°, 300°$.

Without the assistance of a calculator solve $\tan x = -\dfrac{1}{\sqrt{3}}$ for $0 \le x \le 2\pi$.

**Solution**

With the tangent being negative, solutions must lie in the 2nd and 4th quadrants.

From our exact values we know that $\tan \dfrac{\pi}{6} = \dfrac{1}{\sqrt{3}}$.

Thus the solutions make $\dfrac{\pi}{6}$ radians with the $x$-axis as shown diagrammatically on the right.

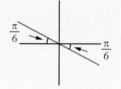

Using this diagram to obtain solutions in the required interval gives: $x = \dfrac{5\pi}{6}, \dfrac{11\pi}{6}$.

Notice that in the previous example the solutions were given in radians because we were told in the question that $0 \le x \le 2\pi$ rather than $0 \le x \le 360°$.

## Using the solve facility on a calculator

The equations of the above examples can be solved using the solve facility available on some calculators (with the calculator set to degrees or radians as appropriate for each question). Note that in the display on the right, the required interval in which we require solutions, is given.

If no interval is stated the calculator will give a general solution involving some constant which, when suitable integer values are substituted for the constant, solutions can be obtained for any required interval. Questions requiring general solutions are not included in this text and we will only encounter equations for which a required interval is stated.

$$\text{solve}\left(\sin(x) = \dfrac{-\sqrt{3}}{2}, x \,\middle|\, 0 \le x \le 360\right)$$
$$x = 240 \text{ or } x = 300$$
$$\text{solve}\left(\tan(x) = \dfrac{-1}{\sqrt{3}}, x \,\middle|\, 0 \le x \le 2 \cdot \pi\right)$$
$$x = \dfrac{5 \cdot \pi}{6} \text{ or } x = \dfrac{11 \cdot \pi}{6}$$

The following examples are solved without using the solve facility. Make sure that you can demonstrate your ability to solve trigonometric equations both with and without a calculator if required to do so.

## EXAMPLE 13

Given that one solution to the equation $\cos x = 0.2$ is, correct to one decimal place, $x = 78.5°$, determine any other solutions the equation has for $-180° \leq x \leq 180°$, giving answers correct to one decimal place.

### Solution

With the cosine of $x$ being positive we know that solutions must lie in the 1st and 4th quadrants.

Hence for $-180° \leq x \leq 180°$ the other solution to the equation is $x = -78.5°$ (correct to one decimal place).

## EXAMPLE 14

Use the information that $\sin 36.9° = 0.6$ to determine all solutions to $5 \sin x = 3$ in the interval $0 \leq x \leq 720°$.

### Solution

$$5 \sin x = 3,$$
$$\therefore \quad \sin x = \frac{3}{5}$$
$$= 0.6$$

Thus the solutions must lie in the 1st and 2nd quadrants and make $36.9°$ with the $x$-axis as shown diagrammatically on the right.

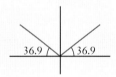

Using this diagram to obtain solutions in the required interval gives: $x = 36.9°, 143.1°, 396.9°, 503.1°$

## EXAMPLE 15

Given that if $x = 1.11$ radians then $\tan x = 2$, solve the equation $\sin x = -2 \cos x$ for $-\pi \leq x \leq \pi$, giving answers in terms of $\pi$ if necessary.

### Solution

$$\sin x = -2 \cos x$$

Dividing both sides by $\cos x$ gives $\dfrac{\sin x}{\cos x} = -2\dfrac{\cos x}{\cos x},$

i.e. $\tan x = -2$

Thus the solutions must lie in the 2nd and 4th quadrants and make 1.11 radians with the $x$-axis as shown diagrammatically on the right.

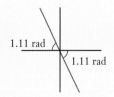

Using this diagram to obtain solutions in the required interval gives:
$x = -1.11$ radians, $(\pi - 1.11)$ radians.

## EXAMPLE 16

Solve $\cos 2x = 0.5$ for $0 \le x \le 2\pi$.

**Solution**

If $\cos 2x = 0.5$ then values of $2x$ must lie in the 1st and 4th quadrants and

make $\dfrac{\pi}{3}$ radians with the $x$-axis.

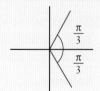

Thus $\quad 2x = \dfrac{\pi}{3}, \dfrac{5\pi}{3}, \dfrac{7\pi}{3}, \dfrac{11\pi}{3},\qquad$ giving $\qquad x = \dfrac{\pi}{6}, \dfrac{5\pi}{6}, \dfrac{7\pi}{6}, \dfrac{11\pi}{6}.$

Notice that in the previous example, to obtain all the solutions for $x$ in the interval 0 to $2\pi$ we had to list values for $2x$ in the interval 0 to $4\pi$. These values for $2x$, when divided by 2, gave values for $x$ in the required interval.

### Note

If we multiply $\sin x$ by itself we could write this as $(\sin x)(\sin x)$ or $(\sin x)^2$. However, to avoid having to write the brackets each time, we write this as $\sin^2 x$. This notation is evident in Example 17.

## EXAMPLE 17

Solve the equation $2 \sin^2 x - 3 \sin x - 2 = 0$ for $0 \le x \le 4\pi$.

**Solution**

The equation is a quadratic in $\sin x$.

To solve the quadratic equation $\qquad 2y^2 - 3y - 2 = 0$

we look for two numbers which add to give $-3$ and multiply to give $-4$. I.e. $+1$ and $-4$.

We then rewrite the equation as $\qquad 2y^2 + 1y - 4y - 2 = 0$

Hence $\qquad\qquad\qquad\qquad y(2y + 1) - 2(2y + 1) = 0$

and so $\qquad\qquad\qquad\qquad\quad (2y + 1)(y - 2) = 0$

Thus, given the equation $\qquad 2 \sin^2 x - 3 \sin x - 2 = 0$

factorising gives $\qquad\qquad (2 \sin x + 1)(\sin x - 2) = 0$

Either $\qquad 2 \sin x + 1 = 0 \qquad\qquad\qquad$ or $\qquad\qquad \sin x - 2 = 0,$

i.e. $\qquad\qquad \sin x = -0.5 \qquad\qquad\qquad$ or $\qquad\qquad \sin x = 2.$

Solutions to $\sin x = -0.5$ must lie in the 3rd and 4th quadrants and make $\dfrac{\pi}{6}$ with the $x$-axis:

From our unit circle definition it follows that $-1 \le \sin x \le 1$.

Thus $\sin x = 2$ has no solution.

Thus $x = \dfrac{7\pi}{6}, \dfrac{11\pi}{6}, \dfrac{19\pi}{6}, \dfrac{23\pi}{6}.$

Thus for $0 \le x \le 4\pi$ the solutions are $x = \dfrac{7\pi}{6}, \dfrac{11\pi}{6}, \dfrac{19\pi}{6}, \dfrac{23\pi}{6}.$

ISBN 9780170390330

## Exercise 8D

**(Do this exercise without using the solve facility of a calculator.)**

Solve the following for $0 \leq x \leq 360°$.

**1** $\cos x = \dfrac{1}{2}$　　　　**2** $\sin x = -\dfrac{1}{2}$　　　　**3** $\tan x = 1$　　　　**4** $\sin x = -\dfrac{1}{\sqrt{2}}$

Solve the following for $0 \leq x \leq 2\pi$.

**5** $\sin x = \dfrac{1}{\sqrt{2}}$　　　　**6** $\cos x = -\dfrac{1}{\sqrt{2}}$　　　　**7** $\tan x = -1$　　　　**8** $\tan x = \sqrt{3}$

Solve the following for $-180° \leq x \leq 180°$.

**9** $\cos x = \dfrac{\sqrt{3}}{2}$　　　　**10** $\sin x = -1$　　　　**11** $\tan x = -\dfrac{1}{\sqrt{3}}$　　　　**12** $\sin x = 0$

Solve the following for $-\pi \leq x \leq \pi$.

**13** $\sin x = \dfrac{\sqrt{3}}{2}$　　　　**14** $\cos x = -\dfrac{1}{2}$　　　　**15** $\sin x = \dfrac{1}{2}$　　　　**16** $\cos x = 0$

**17** For $x$ in radians, one solution of the equation $\tan x = 1.5$ is $x = 0.98$, correct to two decimal places. Hence determine, in terms of $\pi$, any other values of $x$ in the interval $0 \leq x \leq 2\pi$ for which $\tan x = 1.5$.

**18** Use the information that $\cos 63.9° = 0.44$ to determine solutions to
$$11 + 25 \cos x = 0$$
in the interval $-180° \leq x \leq 180°$.

Solve the following for $x$ in the given interval.

**19** $\tan 2x = \dfrac{1}{\sqrt{3}}$ for $0 \leq x \leq 180°$

**20** $\cos 4x = \dfrac{\sqrt{3}}{2}$ for $0 \leq x \leq \pi$

**21** $\sin 3x = \dfrac{1}{2}$ for $-90° \leq x \leq 90°$

**22** $2\sqrt{3} \sin 2x = 3$ for $0 \leq x \leq 2\pi$

**23** $2 \cos 3x + \sqrt{3} = 0$ for $0 \leq x \leq 2\pi$

**24** $(\sin x + 1)(2 \sin x - 1) = 0$ for $0 \leq x \leq 2\pi$

**25** $\sin^2 x = \dfrac{1}{2}$ for $0 \leq x \leq 360°$

**26** $4 \cos^2 x - 3 = 0$ for $-\pi \leq x \leq \pi$

**27** $(\sin x)(2 \cos x - 1) = 0$ for $-180° \leq x \leq 180°$　　**28** Solve $2 \cos^2 x + \cos x - 1 = 0$ for $-\pi \leq x \leq \pi$

**29** Solve $\sin\left(x + \dfrac{\pi}{3}\right) = \dfrac{1}{\sqrt{2}}$ for $0 \leq x \leq 2\pi$.

# The Pythagorean identity

Consider some general point P lying on the unit circle and with coordinates $(a, b)$, as shown in the diagram.

Applying the theorem of Pythagoras to the triangle shown we obtain the result:

$$b^2 + a^2 = 1 \qquad [1]$$

From our unit circle definition of sine and cosine it follows that $a = \cos \theta$ and $b = \sin \theta$.

Substituting these facts into [1] we obtain the Pythagorean identity:

$$\sin^2 \theta + \cos^2 \theta = 1$$

We call this an **identity** because the left hand side, $\sin^2 \theta + \cos^2 \theta$, equals the right hand side, 1, for **all** values of $\theta$.

For example    if $\theta = 10°$,    $\sin^2 10° + \cos^2 10° = 1$ (by calculator),

if $\theta = 30°$,    $\sin^2 30° + \cos^2 30° = \left(\dfrac{1}{2}\right)^2 + \left(\dfrac{\sqrt{3}}{2}\right)^2$

$$= \frac{1}{4} + \frac{3}{4}$$

$$= 1,$$

if $\theta = 45°$,    $\sin^2 45° + \cos^2 45° = \left(\dfrac{1}{\sqrt{2}}\right)^2 + \left(\dfrac{1}{\sqrt{2}}\right)^2$

$$= \frac{1}{2} + \frac{1}{2}$$

$$= 1,$$

if $\theta = 125°$,    $\sin^2 125° + \cos^2 125° = 1$ (by calculator), etc.

This should be compared with an equation which is true only for certain values of $\theta$. For example, the equation $2 \sin \theta = 1$ is true for certain values of $\theta$, e.g. $30°$, $150°$, $390°$ etc., but is not true for all values of $\theta$, e.g. $2 \sin 10° \neq 1$.

In some texts the symbol $\equiv$ is used for an identity.

For example    $\sin^2 \theta + \cos^2 \theta \equiv 1$,   an identity,

but    $2 \sin \theta = 1$,   an equation.

The Pythagorean identity can be used to help us solve some trigonometric equations, as the next example demonstrates.

ISBN 9780170390330

## EXAMPLE 18

Solve $2 \cos^2 \theta + \sin \theta = 2$ for $0 \le \theta \le 360°$.

**Solution**

$$2 \cos^2 \theta + \sin \theta = 2$$

From $\sin^2 \theta + \cos^2 \theta = 1$ it follows that $\cos^2 \theta = 1 - \sin^2 \theta$.

Substituting this expression for $\cos^2 \theta$ into the equation we obtain a quadratic in $\sin \theta$:

$$2 (1 - \sin^2 \theta) + \sin \theta = 2,$$

i.e.
$$2 - 2 \sin^2 \theta + \sin \theta = 2$$

$$\sin \theta - 2 \sin^2 \theta = 0$$

$$\sin \theta (1 - 2 \sin \theta) = 0$$

$\therefore$ either $\qquad \sin \theta = 0 \qquad\qquad$ or $\qquad\qquad 1 - 2 \sin \theta = 0,$

$$\sin \theta = 0.5$$

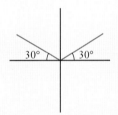

$\theta = 0°, 180°, 360°.$ $\qquad\qquad\qquad\qquad \theta = 30°, 150°.$

Thus for $0 \le \theta \le 360°$ the solutions are $0°, 30°, 150°, 180°$ and $360°$.

As with the solving of trigonometric equations earlier in this chapter, the answers to the previous example can be obtained using the ability of some calculators to solve equations.

Again this facility is very useful but also make sure you can demonstrate your ability to solve 'trig equations' without a calculator when required to do so.

solve$(2 \cdot (\cos(x))^2 + \sin(x) = 2, x) \,|\, 0 \le x \le 360$

$x = 0$ or $x = 30$ or $x = 150$ or $x = 180$ or $x = 360$

## Exercise 8E

Solve the following equations for the given interval but first note the following:

- Whilst you should solve each equation **without** the assistance of a calculator you may find the information in the display below of use for some of them.
- Give exact answers where possible but when rounding is needed give answers correct to one decimal place.
- Not all of the equations require the Pythagorean identity to be used. You must decide whether its use is appropriate.

**1** $\sin x = \dfrac{1}{4}$      for $-180° \leq x \leq 180°$

**2** $\sin^2 x = \dfrac{1}{4}$      for $-\pi \leq x \leq \pi$

**3** $\sin x = \sin^2 x + \cos^2 x$      for $0 \leq x \leq 2\pi$

**4** $(2 \sin x - 1) \cos x = 0$      for $0 \leq x \leq 2\pi$

**5** $\sin x + 2 \sin^2 x = 0$      for $0 \leq x \leq 360°$

**6** $(2 \cos x + 1)(5 \sin x - 1) = 0$      for $0 \leq x \leq 360°$

**7** $8 \sin^2 x + 4 \cos^2 x = 7$      for $0 \leq x \leq 2\pi$

**8** $\tan^2 x + \tan x = 2$      for $-180° \leq x \leq 180°$

**9** $5 - 4 \cos x = 4 \sin^2 x$      for $-90° \leq x \leq 90°$

**10** $3 = 2 \cos^2 x + 3 \sin x$      for $0 \leq x \leq 4\pi$.

```
solve(sin(x) = 0.25, x) | 0 ≤ x ≤ 90°
                        {x = 14.47751219}
solve(sin(x) = 0.2, x) | 0 ≤ x ≤ 90°
                        {x = 11.53695903}
solve(tan(x) = 2, x) | 0 ≤ x ≤ 90°
                        {x = 63.43494882}
factor(2·y²–3·y + 1)
                        (y – 1)·(2·y – 1)
factor(4·y²–4·y + 1)
                        (2·y – 1)²
```

# Angle sum and angle difference

Is the statement $\cos(A - B) = \cos A - \cos B$ true for *all* values of $A$ and $B$? I.e. is the statement an identity? We can demonstrate that **it is not an identity** by considering some values for $A$ and $B$.

For example:  If $A = 90°$ and $B = 30°$ then $\cos(A - B) = \cos(90° - 30°)$

$$= \cos 60°$$

$$= 0.5.$$

But

$$\cos A - \cos B = \cos 90° - \cos 30°$$

$$= 0 - \frac{\sqrt{3}}{2}$$

$$\neq 0.5.$$

Thus $\cos(A - B) \neq \cos A - \cos B$ for these values of $A$ and $B$.

Having established that $\cos(A - B)$ is not the same as $\cos A - \cos B$ can we find an expression that $\cos(A - B)$ is the same as?

Consider the points P and Q lying on the unit circle as shown in the diagram on the right. From our unit circle definition of sine and cosine the coordinates of P and Q will be as shown.

In an earlier chapter we saw that the length of the line joining two points could be found by determining

$$\sqrt{\left(\text{change in the } x\text{-coordinates}\right)^2 + \left(\text{change in the } y\text{-coordinates}\right)^2}$$

Thus $PQ = \sqrt{\left(\cos A - \cos B\right)^2 + \left(\sin A - \sin B\right)^2}$

$$= \sqrt{\cos^2 A - 2\cos A \cos B + \cos^2 B + \sin^2 A - 2\sin A \sin B + \sin^2 B}$$

$$= \sqrt{\cos^2 A + \sin^2 A + \cos^2 B + \sin^2 B - 2\cos A \cos B - 2\sin A \sin B}$$

$$= \sqrt{1 + 1 - 2(\cos A \cos B + \sin A \sin B)}$$

$$= \sqrt{2 - 2(\cos A \cos B + \sin A \sin B)} \qquad \text{[I]}$$

However, if instead we apply the cosine rule to triangle OPQ:

$$PQ = \sqrt{1^2 + 1^2 - 2(1)(1)\cos(A - B)}$$

$$= \sqrt{2 - 2\cos(A - B)} \qquad \text{[II]}$$

Comparing [I] and [II] we see that $\cos(A - B) = \cos A \cos B + \sin A \sin B$.

$$\cos (A - B) = \cos A \cos B + \sin A \sin B \qquad [1]$$

Replacing $B$ by $(-B)$, and remembering that $\cos (-B) = \cos B$ and $\sin (-B) = -\sin B$, it follows that
$$\cos (A - (-B)) = \cos A \cos (-B) + \sin A \sin (-B)$$
$$= \cos A \cos B - \sin A \sin B$$

i.e.
$$\cos (A + B) = \cos A \cos B - \sin A \sin B \qquad [2]$$

From [1],
$$\cos \left( \frac{\pi}{2} - \theta \right) = \cos \frac{\pi}{2} \cos \theta + \sin \frac{\pi}{2} \sin \theta$$
$$= (0) \cos \theta + (1) \sin \theta$$
$$= \sin \theta$$

Replacing $\frac{\pi}{2} - \theta$ by $\phi$ (and hence $\theta$ by $\frac{\pi}{2} - \phi$) it follows that $\cos \phi = \sin \left( \frac{\pi}{2} - \phi \right)$

Thus
$$\cos \left( \frac{\pi}{2} - A \right) = \sin A$$

and
$$\sin \left( \frac{\pi}{2} - A \right) = \cos A$$

(These identities are sometimes referred to as the trigonometric properties of *complementarity*.)

Note
- These facts regarding $A$ and $(90° - A)$ come as no surprise if we remember that this trigonometry for angles of any size must not contradict our initial ideas regarding the trigonometry of right triangles.

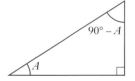

- We can now use these facts to determine expansions for $\sin (A + B)$ and for $\sin (A - B)$.

$$\sin (A - B) = \cos [90° - (A - B)]$$
$$= \cos [90° - A + B]$$
$$= \cos (90° - A) \cos B - \sin (90° - A) \sin B$$
$$= \sin A \cos B - \cos A \sin B$$

i.e.
$$\sin (A - B) = \sin A \cos B - \cos A \sin B \qquad [3]$$

Replacing $B$ by $(-B)$ in [3] gives:
$$\sin (A - (-B)) = \sin A \cos (-B) - \cos A \sin (-B)$$
$$= \sin A \cos B + \cos A \sin B$$

i.e.
$$\sin (A + B) = \sin A \cos B + \cos A \sin B \qquad [4]$$

ISBN 9780170390330

The identities [1], [2], [3] and [4] can be summarised as follows.

$$\sin(A \pm B) = \sin A \cos B \pm \cos A \sin B$$

$$\cos(A \pm B) = \cos A \cos B \mp \sin A \sin B$$

From the previous results it follows that:

$$\tan(A \pm B) = \frac{\sin(A \pm B)}{\cos(A \pm B)}$$

$$= \frac{\sin A \cos B \pm \cos A \sin B}{\cos A \cos B \mp \sin A \sin B}$$

$$= \frac{\dfrac{\sin A \cos B}{\cos A \cos B} \pm \dfrac{\cos A \sin B}{\cos A \cos B}}{\dfrac{\cos A \cos B}{\cos A \cos B} \mp \dfrac{\sin A \sin B}{\cos A \cos B}}$$

$$= \frac{\tan A \pm \tan B}{1 \mp \tan A \tan B}$$

$$\tan(A \pm B) = \frac{\tan A \pm \tan B}{1 \mp \tan A \tan B}$$

## EXAMPLE 19

Determine an exact value for sin 15°.

### Solution

$$\sin 15° = \sin(45° - 30°)$$

$$= \sin 45° \cos 30° - \cos 45° \sin 30°$$

$$= \frac{1}{\sqrt{2}} \frac{\sqrt{3}}{2} - \frac{1}{\sqrt{2}} \frac{1}{2}$$

$$= \frac{\sqrt{3} - 1}{2\sqrt{2}} \text{ or, with a rational denominator, } \frac{\sqrt{2}(\sqrt{3} - 1)}{4}.$$

## EXAMPLE 20

$A$ and $B$ are acute angles with $\cos A = \dfrac{5}{13}$ and $\sin B = \dfrac{24}{25}$.

Find the exact value of $\sin (A + B)$.

**Solution**

If $\cos A = \dfrac{5}{13}$ then $\sin A = \dfrac{12}{13}$ (see diagram).

If $\sin B = \dfrac{24}{25}$ then $\cos B = \dfrac{7}{25}$ (see diagram).

Thus $\sin (A + B) = \sin A \cos B + \cos A \sin B$

$$= \frac{12}{13}\frac{7}{25} + \frac{5}{13}\frac{24}{25}$$

$$= \frac{204}{325}$$

## EXAMPLE 21

Solve $\cos\left(x + \dfrac{\pi}{4}\right) = \sqrt{2}\,\cos x$, for $-2\pi \le x \le 2\pi$.

**Solution**

$$\cos\left(x + \frac{\pi}{4}\right) = \sqrt{2}\cos x$$

$\therefore$ $$\cos x \cos \frac{\pi}{4} - \sin x \sin \frac{\pi}{4} = \sqrt{2}\cos x$$

$$\cos x \frac{1}{\sqrt{2}} - \sin x \frac{1}{\sqrt{2}} = \sqrt{2}\cos x$$

($\times$ by $\sqrt{2}$) $$\cos x - \sin x = 2\cos x$$

$$-\sin x = \cos x$$

($\div$ by $-\cos x$) $$\tan x = -1$$

Thus for $-2\pi \le x \le 2\pi$ the solutions are $-\dfrac{5\pi}{4}, -\dfrac{\pi}{4}, \dfrac{3\pi}{4}, \dfrac{7\pi}{4}$.

ISBN 9780170390330

## Exercise 8F

Simplify each of the following.

**1** $\sin 2x \cos x + \cos 2x \sin x$

**2** $\cos 3x \cos x + \sin 3x \sin x$

**3** $\sin 5x \cos x - \cos 5x \sin x$

**4** $\cos 7x \cos x - \sin 7x \sin x$

Use the formulae for $\sin (A \pm B)$, $\cos (A \pm B)$ and $\tan (A \pm B)$ to determine exact values for each of the following.

**5** $\cos 15°$

**6** $\tan 15°$

**7** $\sin 75°$

**8** $\cos 75°$

**9** $\tan 75°$

**10** If $2 \sin (\theta + 45°) = a \sin \theta + b \cos \theta$, find $a$ and $b$.

**11** If $8 \cos \left( \theta - \dfrac{\pi}{3} \right) = c \sin \theta + d \cos \theta$, find $c$ and $d$.

**12** If $4 \cos (\theta + 30°) = e \cos \theta + f \sin \theta$, find $e$ and $f$.

**13** If $\tan A = 5\sqrt{3}$ and $\tan B = -\dfrac{\sqrt{3}}{4}$, find (without a calculator) the value of $\tan (A + B)$.

If $\pi \le (A + B) \le 2\pi$ determine $(A + B)$.

**14** $A$ and $B$ are acute angles with $\sin A = \dfrac{4}{5}$ and $\cos B = \dfrac{5}{13}$. Find the following as exact values.

   **a**   $\sin (A + B)$

   **b**   $\cos (A - B)$

**15** $D$ and $E$ are acute angles with $\sin D = \dfrac{7}{25}$ and $\sin E = \dfrac{3}{5}$. Find the following as exact values.

   **a**   $\sin (D - E)$

   **b**   $\cos (D + E)$

**16** Use the expansion of $\sin (A + B)$ to prove that $\sin \left( x + \dfrac{\pi}{2} \right) = \cos x$.

**17** Use the expansion of $\sin (A \pm B)$ to prove

   **a**   $\sin (x + 2\pi) = \sin x$

   **b**   $\sin (x - 2\pi) = \sin x$

**18** Use the expansion of $\cos (A + B)$ to prove that $\cos (x + 2\pi) = \cos x$.

**19** Use the expansion of $\tan (A + B)$ to prove that $\tan (x + \pi) = \tan x$.

**20** By writing $\tan (-x)$ as $\tan (0 - x)$, use the $\tan (A - B)$ expansion to prove that $\tan (-x) = - \tan x$.

**21** $A$ and $B$ are both obtuse angles such that $\sin A = \dfrac{5}{13}$ and $\tan B = -\dfrac{3}{4}$. Find exact values for

   **a**   $\sin (A + B)$

   **b**   $\cos (A - B)$

   **c**   $\tan (A + B)$

Solve the following equations for the given interval.

**22** $\sin x \cos \dfrac{\pi}{6} + \cos x \sin \dfrac{\pi}{6} = \dfrac{1}{\sqrt{2}}$ for $0 \le x \le 2\pi$.

**23** $\cos x \cos 20° + \sin x \sin 20° = \dfrac{1}{2}$ for $0 \le x \le 360°$.

**24** $\sin x \cos 70° + \cos x \sin 70° = 0.5$ for $-180° \le x \le 180°$.

**25** $\sin (x + 30°) = \cos x$ for $0 \le x \le 360°$.

## Alternating currents

An electrical current is a flow of electrical charge. In wires this electrical charge is carried by electrons. Batteries produce a steady, one directional, flow of electrons called a direct current (DC). If instead the electrons repeatedly move one way and then the other their alternating flow will still result in a flow of electrical charge, and hence a current. This is called an alternating current or AC. Many household electrical devices simply require an electrical current, they do not require the flow of electrons to always be in a certain direction. Hence for such devices an alternating current is suitable.

An alternating current is produced by an alternating voltage, which is the form of voltage that the electrical supply companies supply to most homes and businesses.

These alternating voltages are sinusoidal in nature.

Let us suppose that the voltage, $V$ Volts, at time $t$ seconds is as shown below:

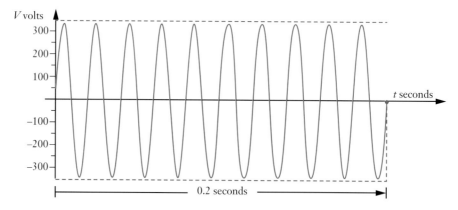

What is the amplitude and period of this graph?

Write the equation of the graph in the form $V = a \sin bt$, with $b$ in terms of $\pi$.

(Hint: Remember that $y = a \sin bx$ has a period of $\dfrac{2\pi}{b}$.)

ISBN 9780170390330

## Average weekly temperatures

Records taken of the average weekly temperatures in a particular location gave rise to the following graph:

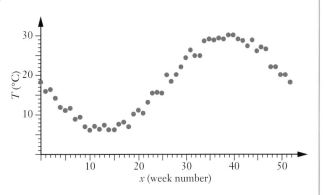

For how many of the weeks did the average weekly temperature exceed 25°C?

Suggest a suitable equation for this data. (Remember, $y = a \sin bx$ has a period of $\dfrac{2\pi}{b}$.)

## Tidal motion

The tide height was measured at a high tide and each hour thereafter for 25 hours. The data collected gave rise to the following table:

| Hours from hightide (t) | 0 | 1 | 2 | 3 | 4 | 5 | 6 | 7 | 8 |
|---|---|---|---|---|---|---|---|---|---|
| Height (h metres) | 12.60 | 12.03 | 10.38 | 8.17 | 5.91 | 4.12 | 3.25 | 3.54 | 4.92 |

| Hours from hightide (t) | 9 | 10 | 11 | 12 | 13 | 14 | 15 | 16 | 17 |
|---|---|---|---|---|---|---|---|---|---|
| Height (h metres) | 7.02 | 9.36 | 11.31 | 12.46 | 12.46 | 11.31 | 9.38 | 7.01 | 4.92 |

| Hours from hightide (t) | 18 | 19 | 20 | 21 | 22 | 23 | 24 | 25 |
|---|---|---|---|---|---|---|---|---|
| Height (h metres) | 3.51 | 3.24 | 4.08 | 5.90 | 8.21 | 10.42 | 12.03 | 12.60 |

Plot a graph of these figures and draw a smooth curve that seems to best fit the facts.

Suggest a rule for your 'best fit' line.

Use your rule to determine the values of $t$, for $0 \le t \le 25$, between which the tide height is at least 5 metres.

Investigate whether or not your calculator can determine a line of best fit for the given data (look for a *regression* facility). If you are able to obtain this line of best fit from your calculator compare it to the rule you determined.

iStock.com/alfonso_c_orive

# Miscellaneous exercise eight

This miscellaneous exercise may include questions involving the work of this chapter, the work of any previous chapters, and the ideas mentioned in the Preliminary work section at the beginning of the book.

State the amplitude and period of each of the following sinusoidal functions.
(Assume radian measure used.)

**1** $y = 5 \sin x$          **2** $y = 7 \sin x$          **3** $y = -3 \sin x$

**4** $y = \sin 2x$          **5** $y = \sin 3x$          **6** $y = \sin 0.5x$

**7** $y = -3 \sin 4x$        **8** $y = 4 \sin 5x$          **9** $y = 2 \sin \pi x$

**10** Copy and complete the following table (without the assistance of a calculator), placing appropriate **exact** values in each empty cell and all denominators rational.

| θ | $-\dfrac{3\pi}{4}$ | $-\dfrac{2\pi}{3}$ | $\dfrac{\pi}{6}$ | $\dfrac{\pi}{4}$ | $\dfrac{4\pi}{3}$ | $\dfrac{7\pi}{3}$ | $\dfrac{9\pi}{4}$ | $11\pi$ |
|---|---|---|---|---|---|---|---|---|
| Sin θ | | | | | | | | |
| Cos θ | | | | | | | | |
| Tan θ | | | | | | | | |

**11** For each of the following pairs of lines determine whether they are parallel, perpendicular or neither of these.

**a** $\begin{cases} y = 3x - a \\ y = x - b \end{cases}$      **b** $\begin{cases} y = 0.5x + c \\ 2y = x + d \end{cases}$      **c** $\begin{cases} 2y = x + e \\ y = -2x + f \end{cases}$

**12** For the triangle shown on the right find $x$, using

    **a** the cosine rule and the solve facility of your calculator

    **b** the sine rule twice

giving your answer correct to one decimal place each time.

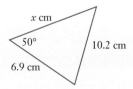

**13** A triangle has sides of length 27 cm, 33 cm and 55 cm. Find the size of the smallest angle of the triangle, giving your answer to the nearest degree.

**14** For each of the following, without using a calculator, write the coordinates of the points where the graph cuts or touches the $x$-axis. (Then check your answers with your calculator if you wish.)

    **a** $y = (x - 2)(x - 3)(x + 2)(x + 7)$       **b** $y = x(x - 2)(x + 3)(x - 4)$

    **c** $y = (x - 2)(x - 3)(x + 3)^2$          **d** $y = (x - 2)^4$

    **e** $y = (x - 7)(2x^2 - 3x + 2)$          **f** $y = (x^2 - x - 30)(4x^2 - 8x - 21)$

**15** All of the functions $f_1$ to $f_{12}$ have their graphs shown below. Find the values of $k_1$ to $k_{26}$ given that they are all non-zero constants between $-50$ and $50$.

$f_1 : y = k_1 x$

$f_2 : x + y + k_2 = 0$

$f_3 : y = k_3 x^2$

$f_4 : y = (k_4 x)^2$

$f_5 : y = k_5 x^2 + k_6 x + k_7$

$f_6 : xy = k_8$

$f_7 : y = k_9 (x + k_{10})^3 + k_{11}$

$f_8 : y = k_{12} (x + k_{13})^2 + k_{14}$

$f_9 : y = k_{15} x^3 + k_{16} x^2 + k_{17} x + k_{18}$

$f_{10} : y = k_{19} \sin k_{20} x$

$f_{11} : y = k_{21} \cos k_{22}(x - k_{23})$

$f_{12} : y = k_{24} + k_{25} \sin k_{26} x$

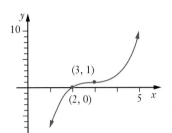

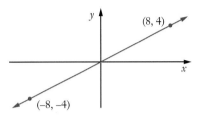

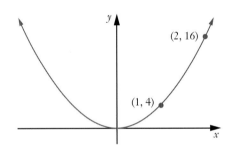

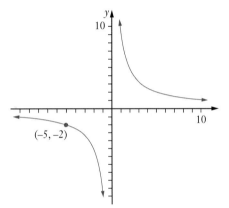

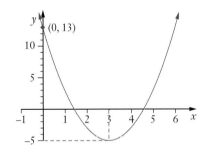

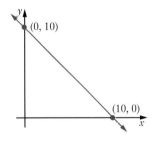

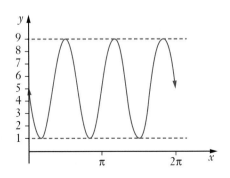

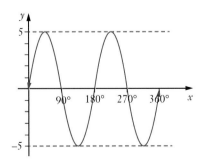

**16** Given that $2x^3 + x^2 - 22x + 24 = (x - 2)(ax^2 + bx + c)$:

   **a** determine the value of $a$ and the value of $c$ by inspection.

   **b** with your answers from part **a** in place, expand $(x - 2)(ax^2 + bx + c)$ and hence determine $b$.

   **c** find the coordinates of the point(s) where the graph of $y = 2x^3 + x^2 - 22x + 24$ cuts the $x$-axis.

**17** The function $y = f(x)$, shown graphed on the right, has a maximum turning point at $(-1, 21)$ and a minimum turning point at $(3, -11)$.

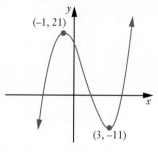

State the coordinates of the maximum and minimum turning points of each of the following functions.

   **a**  $y = f(x) + 5$          **b**  $y = f(x) - 5$

   **c**  $y = f(-x)$            **d**  $y = -f(x)$

   **e**  $y = 3f(x)$            **f**  $y = f(2x)$

**18** **Work through this question without the assistance of a graphic calculator.**

   For the graph of          $y = (x + 1)(x - 2)(x - 5)$

   **a** find the coordinates of any point(s) where the curve cuts the $x$-axis.

   **b** find the coordinates of any point(s) where the curve cuts the $y$-axis.

   **c** if the point A$(1, a)$ lies on the curve determine the value of $a$.

   **d** if the point B$(3, b)$ lies on the curve determine the value of $b$.

   **e** if the point C$(4, c)$ lies on the curve determine the value of $c$.

   For the graph of          $y = (x - 3)^2 - 4$

   **f** find the coordinates of the turning point and state whether it is a maximum point or a minimum point.

   **g** find the coordinates of any point(s) where the curve cuts the $y$-axis.

   **h** if the point D$(5, d)$ lies on the curve determine the value of $d$.

   **i** use the information from parts **a** to **h** above to produce a single sketch showing the two functions and hence estimate solutions to the equation:
$$(x + 1)(x - 2)(x - 5) = (x - 3)^2 - 4.$$

**19** Determine the area of the shaded region shown on the right given that the circle has a radius of 10 cm and AB is of length 16 cm. Give your answer correct to the nearest 0.1 cm$^2$.

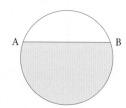

# Sets and probability

**Before commencing this chapter the reader is advised to spend a few minutes re-reading the brief sections on sets and probability in the *Preliminary work* at the beginning of this book.**

# Sets and probability – basic ideas

The next exercise provides you with some practice in the sets and probability ideas referred to in the *Preliminary work*.

Set operations

## Exercise 9A

### Probability

Dartboard probabilities

**1** A normal fair six sided die is rolled once. With P(X) meaning the probability of event X occurring, determine

   **a** P(an even number)

   **b** P(an odd number)

   **c** P(a prime number)    Remember: A prime has exactly 2 factors

   **d** P(a number greater than four)

   **e** P(a number not less than three)

Venn diagrams

**2** Two fair dice are rolled once and the numbers on the uppermost faces are added together. The table on the right shows the 36 equally likely outcomes.

With P(X) meaning the probability of event X occurring, determine

   **a** P(an even total)

   **b** P(an odd total)

   **c** P(a prime total)

   **d** P(a total of 11)

   **e** P(a total that is greater than 8)

   **f** P(a total that is not greater than 8)

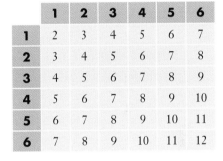

|   | 1 | 2 | 3 | 4 | 5 | 6 |
|---|---|---|---|---|---|---|
| **1** | 2 | 3 | 4 | 5 | 6 | 7 |
| **2** | 3 | 4 | 5 | 6 | 7 | 8 |
| **3** | 4 | 5 | 6 | 7 | 8 | 9 |
| **4** | 5 | 6 | 7 | 8 | 9 | 10 |
| **5** | 6 | 7 | 8 | 9 | 10 | 11 |
| **6** | 7 | 8 | 9 | 10 | 11 | 12 |

Venn diagrams

**3** A particular event has just three possible outcomes A, B and C. Only one of these outcomes can occur each time the event happens and P(A) = 0.5 and P(B) = 0.2. If the event occurs once determine

   **a** P(C)        **b** P(not C)        **c** P(not A)        **d** P(not B)

iStock.com/Playwitt

**4** If a coin is flipped three times there are eight equally likely outcomes:

For such an event determine:

**a**  P(HHH)

**b**  P(Two heads and a tail in that order)

**c**  P(Two heads and a tail in any order)

**d**  P(The third flip produces a head)

**e**  P(The third flip produces the only head)

**f**  P(The third flip produces the second head)

**5** A fair die is rolled onto a flat wooden table and when the die has come to rest the five numbers that can be seen are added together.

For one such roll determine:

**a**  P(The total obtained is less than 15)

**b**  P(The total obtained is more than 15)

**c**  P(The total obtained is divisible by 3)

**d**  P(The total obtained is divisible by 5)

**e**  P(The total obtained is divisible by both 3 and 5)

iStock.com/hammerstron

**6** A child makes a spinner which, on 100 spins produces the results:

| Result of spin | 1 | 2 | 3 | 4 | 5 | 6 | 7 |
|---|---|---|---|---|---|---|---|
| Frequency | 12 | 10 | 8 | 22 | 20 | 15 | 13 |

*Based on these figures* what is the probability that on one spin of the spinner the result will be

**a**  2

**b**  6

**c**  even

**d**  > 4

**e**  ≥ 4

**f**  < 3.

**7** The causes of death given for the sixty seven thousand two hundred and forty one Australian males who died in one particular year were as follows:

| Cardiovascular disease | Cancers | Traffic accidents | All others |
|---|---|---|---|
| 21 957 | 22 039 | 1224 | 22 021 |

[Source of data: National Heart Foundation of Australia and The Australian Bureau of Statistics.]

An insurance company uses these figures to determine the probable causes of death amongst Australian males for the following year. What do these figures suggest for the probability of the death of an Australian male being due to

**a**  cardiovascular disease (correct to 3 decimal places)

**b**  a cause other than cancer (correct to 3 decimal places).

**Sets**

**8** From the Venn diagram on the right we see that

$$A = \{1, 2, 3, 4, 5, 6, 7\}.$$

**a** State n(A).

**b** State n(A ∪ B).

**c** State n(U).

**d** State n(A ∩ B).

**e** List the elements of A′.

**f** List the elements of $\overline{B}$.

**g** List the elements of $\overline{A \cup B}$.

**h** List the elements of $\overline{A \cap B}$.

**9** The universal set, U, and the two sets A and B contained within it are such that

$$n(A \cap B) = 6, n(A) = 27, n(B) = 46 \text{ and } n(U) = 70.$$

Determine      **a**   n(A ∪ B)      **b**   $n(\overline{A \cup B})$

**10** Eighty students commenced a particular university course. Of these, 42 had studied Physics in their last year at school and 46 had studied Chemistry in their last year at school. If 25 had studied both of these subjects in their last year at school how many had studied neither?

**11** The Venn diagram on the right shows the number of elements in each of sets A and B and in the Universal set, U, in which A and B are contained.

If n(U) = 72 find the value of *x*.

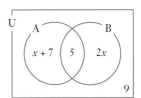

**12** Of the 137 year eleven students in a school the number who had represented the school at volleyball but not athletics was 7 more than the number who had represented the school at both volleyball and athletics.

The number who had represented the school at volleyball but not athletics was twice the number who had represented the school at athletics but not volleyball.

The number who had represented the school at neither volleyball nor athletics was nine times the number who had represented the school at both volleyball and athletics.

How many of the year eleven students had represented the school at both volleyball and athletics?

**13** The Venn diagram on the right shows the number of elements in each of sets A, B and C and in the Universal set, U, in which A, B and C are contained.

If $n(\overline{B \cup C}) = 11$ find      **a**   n(U),

         **b**   n(A ∩ B),

         **c**   n(A ∩ B ∩ C).

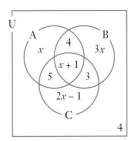

## Probability and sets

In questions **14** to **16**, the notation P(X) is used to mean the probability that an element randomly selected from the universal set U is in set X.

**14** The Venn diagram shows the probabilities of the events A and B occurring. Find

    **a** P(A)     **b** P(B)     **c** P(A ∩ B)

    **d** P(A ∪ B)     **e** P($\overline{A \cap B}$)     **f** P($\overline{A \cup B}$)

**15** The Venn diagram on the right shows how the 19 elements in the universal set U are placed with regards to sets C and D.

    Determine:

    **a** P(C)     **b** P(D)     **c** P(C ∩ D)

    **d** P($\overline{C \cap D}$)     **e** P(C ∩ $\bar{D}$)     **f** P($\bar{C}$ ∩ D)

    **g** P(C ∪ D)     **h** P(C ∪ $\bar{D}$)     **i** P($\bar{C}$ ∪ D)

**16** The Venn diagram on the right shows how the 40 elements in the universal set U are placed with regards to sets X and Y.

    Determine:

    **a** P(X)     **b** P(Y)     **c** P(X ∩ Y)

    **d** P(X ∪ Y)     **e** P($\overline{X \cup Y}$)     **f** P($\overline{X \cap Y}$)

**17** Two events A and B are such that P(A) = 0.3, P(B) = 0.5 and P(A ∪ B) = 0.6, where the notation P(A) means the probability of event A occurring. Determine P(A ∩ B).

**18** If a student is randomly selected from the year 12 students in a particular school the probability of that student being male is 0.52, the probability that they study chemistry is 0.44 and the probability of them being female and not doing chemistry is 0.18.

Determine the probability that the selected student is

    **a** a female studying chemistry     **b** a male not studying chemistry

**19** Twenty dancers from a particular dance school attend the national dance championships. All of the twenty are entered in either solo events or team events or both solo and team. Five of the twenty are not entered in solo events and two are not entered in team events.

    **a** If one of the twenty dancers is randomly selected to represent the school in the opening ceremony what is the probability that this dancer is one who is entered in both solo and team events?

    **b** One of the dancers is injured when competing in the solo event. What is the probability that this dancer is one who is entered in team events?

## Conditional probability

In some situations we may be *given* some extra piece of information, or some additional *condition*, which allows us to reduce the number of possibilities that we need to consider in the sample space. Indeed the last question in the previous exercise is like this.

The information given in part **b** informs us that the dancer we are considering is one who competes in a solo event, and so allows us to consider only the solo dancers.

The following examples and the questions of **Exercise 9B** give further examples of this idea of *conditional probability*.

## EXAMPLE 1

Arika rolls a fair die once and May, who cannot see the result, tries to guess the outcome.

**a**    What is the probability that the result is a 5?

Before May states her guess, Arika announces 'It's an odd number'.

**b**    Now what is the probability that the result is a 5?

### Solution

**a**    There are six equally likely outcomes: 1, 2, 3, 4, 5, 6.

A result of a 5 is one of these 6.        Thus $P(5) = \dfrac{1}{6}$.

**b**    Given the information *It's an odd number* there are now just three equally likely outcomes: 1, 3, 5.

A result of a 5 is one of these 3.        Thus $P(5) = \dfrac{1}{3}$.

## EXAMPLE 2

The Venn diagram indicates the number of people in each of the sets A and B.

Determine the probability that a person chosen at random from the universal set, U, is

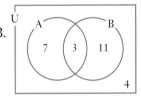

**a**    in set B

**b**    outside of set B

**c**    in set A given they are in set B,

**d**    in set A given they are in A ∪ B.

### Solution

**a**    $P(\text{person is in set B}) = \dfrac{14}{25}$

**b**    $P(\text{outside of set B}) = \dfrac{11}{25}$

**c**    Given that the person is in set B we need only consider the 14 people in B (see diagram).

3 of these 14 are in set A.

Thus    $P(\text{person is in set A given they are in set B}) = \dfrac{3}{14}$.

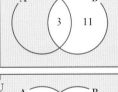

**d**    Given that the person is in A ∪ B we need only consider the 21 people in A ∪ B (see diagram).

10 of these 21 are in set A.

Thus    $P(\text{person is in set A given they are in A} \cup \text{B}) = \dfrac{10}{21}$.

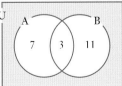

# Notation

In part **c** of the previous example we were asked for the probability of a person being in set A **given** they are in set B. We write the probability of some event X occurring **given** some other condition Y as

$$P(X \mid Y)$$

and we refer to this as the probability of X *given* Y.

Thus in part **c** of the last example:

$$P(\text{person is in set A} \mid \text{they are in set B}) = \frac{3}{14}.$$

Check that you agree with the following statements.

For a single roll of a fair die:

$$P(5 \mid \text{odd number}) = \frac{1}{3},$$

| 1 | 2 | 3 | 4 | 5 | 6 |
|---|---|---|---|---|---|

$$P(5 \mid \text{greater than 4}) = \frac{1}{2},$$

| 1 | 2 | 3 | 4 | 5 | 6 |
|---|---|---|---|---|---|

$$P(5 \mid \text{less than 4}) = 0.$$

| 1 | 2 | 3 | 4 | 5 | 6 |
|---|---|---|---|---|---|

## EXAMPLE 3

A fair octahedral die is rolled once. Determine

**a** P(7)  **b** P(7 | odd number)  **c** P(3 | prime number)

**d** P(not a 2 | a number < 4)  **e** P(odd number | 7)

**Solution**

**a** On an octahedral die there are eight equally likely outcomes: 1, 2, 3, 4, 5, 6, 7, 8.

A result of a 7 is one of these 8 equally likely outcomes.   Thus $P(7) = \dfrac{1}{8}$.

**b** Given that the result is an odd number there are now just four equally likely outcomes: 1, 3, 5, 7. A result of a 7 is one of these 4 equally likely outcomes.

Thus: $$P(7 \mid \text{odd number}) = \frac{1}{4}.$$

**c** Given that the result is a prime number there are just four equally likely outcomes: 2, 3, 5, 7. A result of a 3 is one of these 4 equally likely outcomes.

Thus: $$P(3 \mid \text{prime number}) = \frac{1}{4}.$$

**d** Given that the result is a number less than 4 there are just three equally likely outcomes: 1, 2, 3. Two of these three equally likely outcomes are 'not a 2'.

Thus: $$P(\text{not a 2} \mid \text{a number} < 4) = \frac{2}{3}.$$

**e** Given that the result is a 7 there is just one outcome, a 7, and this is an odd number.

Thus: $$P(\text{odd number} \mid 7) = \frac{1}{1} = 1.$$

ISBN 9780170390330

EXAMPLE 4

The Venn diagram on the right shows the probabilities of events A and B occurring. Determine

**a** $P(A \cap B)$      **b** $P(A|B)$      **c** $P(B|\overline{A})$

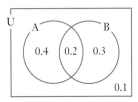

**Solution**

**a** $P(A \cap B) = 0.2$

**b** Given that event B occurs we need only consider that part of the Venn diagram (see unshaded parts on right).

The total under consideration is now 0.5.

0.2 lies in circle A.

Thus $P(A|B) = \dfrac{0.2}{0.5}$

$= 0.4$

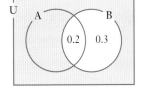

**c** Given that $\overline{A}$ occurs we need only consider that part of the Venn diagram (see unshaded parts on right).

The total under consideration is now 0.4.
0.3 lies in circle B.

Thus $P(B|\overline{A}) = \dfrac{0.3}{0.4}$

$= 0.75$

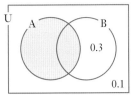

## Exercise 9B

**1** Jack rolls a normal fair die once and Holly, who cannot see the result, tries to guess the outcome.

**a** What is the probability that the result is a 4?

Before Holly states her guess Jack announces 'It's not a six'.

**b** Now what is the probability that the result is a 4?

**2** Leroy tosses two fair coins and Boon attempts to guess the result.

**a** What is the probability that the result is two heads?

Before Boon announces his guess Leroy states 'It's not two tails'.

**b** Now what is the probability that the result is two heads?

     (H)(H)

     (H)(T)

     (T)(H)

     (T)(T)

**3** Leslie tosses two fair normal six sided dice, one red and one blue, and Ranji attempts to guess the total obtained when the number of dots on each uppermost face are added.

**a** What is the probability that the total is eleven?

In fact, Ranji manages to see that the red die lands with a six uppermost.

**b** Now what is the probability that the total is eleven?

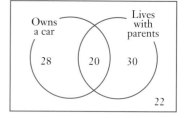

| | | RED DIE | | | | | |
|---|---|---|---|---|---|---|---|
| | | **1** | **2** | **3** | **4** | **5** | **6** |
| **BLUE DIE** | **1** | 2 | 3 | 4 | 5 | 6 | 7 |
| | **2** | 3 | 4 | 5 | 6 | 7 | 8 |
| | **3** | 4 | 5 | 6 | 7 | 8 | 9 |
| | **4** | 5 | 6 | 7 | 8 | 9 | 10 |
| | **5** | 6 | 7 | 8 | 9 | 10 | 11 |
| | **6** | 7 | 8 | 9 | 10 | 11 | 12 |

**4** Mansur randomly selects a card from a normal well shuffled pack of 52 playing cards and Marlena attempts to guess what card has been selected.

Hearts        A♥, 2♥, 3♥, 4♥, 5♥, 6♥, 7♥, 8♥, 9♥, 10♥, J♥, Q♥, K♥

Diamonds     A♦, 2♦, 3♦, 4♦, 5♦, 6♦, 7♦, 8♦, 9♦, 10♦, J♦, Q♦, K♦

Spades       A♠, 2♠, 3♠, 4♠, 5♠, 6♠, 7♠, 8♠, 9♠, 10♠, J♠, Q♠, K♠

Clubs        A♣, 2♣, 3♣, 4♣, 5♣, 6♣, 7♣, 8♣, 9♣, 10♣, J♣, Q♣, K♣

**a** What is the probability that the card is the two of hearts?

**b** If in fact Marlena manages to see that the card is red and is not a Jack, King or Queen what is the probability now that it is the two of hearts?

**5** The Venn diagram on the right displays the results of a survey of 100 university students regarding whether they own a car and if they are living with their parents.

**a** What is the probability that a student chosen at random from this group does own a car but does not live with their parents?

**b** Given that a student chosen from this group does not own a car what is the probability that they live with their parents?

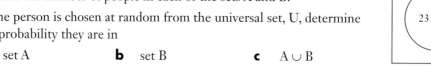

**6** The numbers in the various sections of the Venn diagram on the right indicate the number of people in each of the sets A and B.

If one person is chosen at random from the universal set, U, determine the probability they are in

**a** set A        **b** set B        **c** A ∪ B

**d** A ∩ B      **e** $\bar{A}$        **f** $\bar{B}$

**g** set A given they are in set B       **h** set A given they are not in set B

**7** A fair die is rolled once. Event A is that of the result of the roll being a 6.
Event B is that of the result of the roll being bigger than 4.
Event C is that of the result of the roll being an even number.

Determine

**a** P(A|B)       **b** P(A|C)       **c** P(B|C)

**d** P(B|A)       **e** P(C|A)       **f** P(C|B)

**8** The Venn diagram on the right shows the probabilities of the events A and B occurring. Find

  **a**  P(A)        **b**  P(B)        **c**  P(A ∪ B)

  **d**  P(Ā)        **e**  P(B̄)        **f**  P(A|B)

  **g**  P(B|A)      **h**  P(A|A ∪ B)      **i**  P(B|A ∩ B)

U  A     B   0.6  0.1  0.2   0.1

**9** The Venn diagram on the right shows the probabilities of the events A and B occurring. Find

  **a**  P(A)        **b**  P(B)        **c**  P(B ∪ A)

  **d**  P(A ∩ B)    **e**  P(Ā)        **f**  P(B̄)

  **g**  P(A|B)      **h**  P(B|A)      **i**  P(B|A ∪ B)

U  A     B   $\frac{2}{9}$  $\frac{1}{9}$  $\frac{4}{9}$   $\frac{2}{9}$

**10** A team of two people is to be randomly selected from the following five people:

Alex, Basil, Chris, Deny, Ernie.

There are ten different teams of two that can be made:

| Alex Basil | Alex Chris | Alex Deny | Alex Ernie | Basil Chris | Basil Deny | Basil Ernie | Chris Deny | Chris Ernie | Deny Ernie |
|---|---|---|---|---|---|---|---|---|---|

  **a**  Find the probability that Alex is selected.

  **b**  Find the probability that the team of two consists of Deny and Alex.

  **c**  Given that Deny is selected, what is the probability that Alex is also selected?

  **d**  What is the probability that just one of Deny and Alex is selected?

  **e**  Given that just one of Deny and Alex is selected, what is the probability that Ernie fills the other space on the team?

**11** A fair normal six sided die is rolled once. Determine

  **a**  P(3)                  **b**  P(3|odd number)

  **c**  P(3|even number)        **d**  P(even number|3)

  **e**  P(2|a number < 4)       **f**  P(not a 2|a number < 4)

  **g**  P(a number < 2|a number < 4)    **h**  P(a number ≤ 2|a number < 4)

**12** A fair ten sided die, with faces marked 1 to 10, is rolled once. Determine

  **a**  P(8)                  **b**  P(8|even number)

  **c**  P(8|prime number)       **d**  P(7|prime number)

  **e**  P(8|a number > 4)        **f**  P(not an 8|a number > 4)

  **g**  P(a number > 8|a number ≥ 6)

**13** From a set of cards numbered 1 to 20 one card is selected at random. What is the probability that the card is

  **a**  a seven given that it is less than ten?      **b**  a seven given that it is more than ten?

  **c**  a six given that it is a multiple of three?    **d**  a six given that it is a factor of twelve?

  **e**  an even number given that it is neither four nor ten?

**14** A normal six sided die is rolled twice and the scores obtained are added together.

Event A is that of the first roll giving a 5.
Event B is that of the total being 10.

Determine

**a** $P(B|A)$          **b** $P(A|B)$

Event C is that of the first roll giving a 2.
Event D is that of the total being 6.

Determine

**c** $P(D|C)$          **d** $P(C|D)$

Event E is that of the first roll giving an odd number.
Event F is that of the total being 5.

Determine

**e** $P(F|E)$          **f** $P(E|F)$

| | | SECOND ROLL | | | | | |
|---|---|---|---|---|---|---|---|
| | | **1** | **2** | **3** | **4** | **5** | **6** |
| FIRST ROLL | **1** | 2 | 3 | 4 | 5 | 6 | 7 |
| | **2** | 3 | 4 | 5 | 6 | 7 | 8 |
| | **3** | 4 | 5 | 6 | 7 | 8 | 9 |
| | **4** | 5 | 6 | 7 | 8 | 9 | 10 |
| | **5** | 6 | 7 | 8 | 9 | 10 | 11 |
| | **6** | 7 | 8 | 9 | 10 | 11 | 12 |

**15** A fair icosahedral die (20 faces numbered 1 to 20) is rolled once.

Determine

**a** $P(3)$        **b** $P(\text{even})$        **c** $P(\text{prime})$

**d** $P(\text{multiple of 3})$        **e** $P(\text{factor of 12})$        **f** $P(5|\text{an odd number})$

**g** $P(3|<6)$        **h** $P(15|>9)$        **i** $P(9|\text{a multiple of 3})$

**j** $P(\text{a multiple of 3}|\text{a factor of 12})$        **k** $P(\text{multiple of 3}|>15)$        **l** $P(\text{multiple of 3}|\geq 15)$

**16** The twenty four 3-digit numbers that can be formed using the digits 1, 2, 3, and 5, with each digit being used only once in each number are as follows:

| | | | | | |
|---|---|---|---|---|---|
| 123 | 132 | 213 | 231 | 312 | 321 |
| 235 | 253 | 325 | 352 | 523 | 532 |
| 135 | 153 | 315 | 351 | 513 | 531 |
| 125 | 152 | 215 | 251 | 512 | 521 |

One of these twenty four numbers is to be selected at random.
Event A is that of the selected number being bigger than 300.
Event B is that of the selected number being bigger than 400.

Determine

**a** $P(A)$        **b** $P(B)$        **c** $P(A|B)$

**d** $P(B|A)$        **e** $P(A|\overline{B})$        **f** $P(\overline{A}|B)$

**17** (Hint: A little thought could save you time with this question.)

A fair coin is tossed four times. What is the probability that the result of the next throw, i.e. the fifth throw, will be a head given that

**a** the fourth throw resulted in a head?

**b** the third and the fourth throws resulted in heads?

---

**RESEARCH**

Research and write a brief report on *The gambler's fallacy*.

---

# Tree diagrams

As included in the brief mention of probability in the *Preliminary work* at the start of this book, another useful form of presentation when finding probabilities is a tree diagram. The next example demonstrates the use of this method of displaying the equally likely outcomes of an event. The last two parts of the example also involve conditional probability.

## EXAMPLE 5

Envelope I contains the three letters: A, B and C.

Envelope II contains the three letters: B, C and D.

A two letter 'word' is formed, the first letter being randomly chosen from envelope I and the second letter randomly chosen from envelope II.

The tree diagram on the right shows the 9 equally likely outcomes.

Determine the probability that

**a**  exactly one letter B is chosen

**b**  the letters chosen are the same

**c**  the first letter chosen is an A given that the two letters are not the same

**d**  the second letter chosen is a B given the two letters are the same.

| Envelope I | Envelope II | Outcome |
|---|---|---|
| A | B | AB |
|  | C | AC |
|  | D | AD |
| B | B | BB |
|  | C | BC |
|  | D | BD |
| C | B | CB |
|  | C | CC |
|  | D | CD |

## Solution

**a**  Four of the nine equally likely outcomes contain exactly one B.

$$P(\text{Exactly one B}) = \frac{4}{9}$$

**b**  In two of the nine equally likely outcomes the letters are the same.

$$P(\text{Same letters}) = \frac{2}{9}$$

**c**  Given that the two letters are not the same we need only consider those outcomes shown arrowed in the tree diagram on the right.

Three of these seven outcomes start with A.

$$P(\text{Start with A}\,|\,\text{not same letters}) = \frac{3}{7}$$

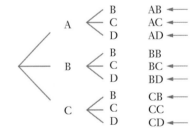

**d**  Given that the two letters are the same we only consider those outcomes shown arrowed in the tree diagram on the right.

$$P(\text{Second letter B}\,|\,\text{same letters}) = \frac{1}{2}$$

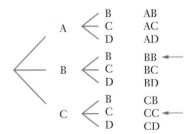

**Exercise 9C**

**1** Two marbles are drawn at random from a bag containing four marbles – 2 red, 1 blue and 1 green. After the first marble is drawn and its colour noted, it is returned to the bag before the second marble is drawn. Construct a suitable tree diagram and hence determine the following probabilities:

   **a**  P(two reds | first one red)
   **b**  P(two reds | both same colour)
   **c**  P(two different | 2nd one green)
   **d**  P(no reds | 2nd one green)

**2** Two marbles are drawn at random from a bag containing four marbles – 2 red, 1 blue and 1 green. After the first marble is drawn and its colour noted, it is **not** returned to the bag. Construct a suitable tree diagram and hence determine the following probabilities:

   **a**  P(two reds | first red)
   **b**  P(one green | first not blue)
   **c**  P(first blue | second red)
   **d**  P(second red | first blue)

**3** Laurie, Rob and Steven play two rounds of a game of cards. The game has no draws, ties or stalemates so each round is won by one of the three people and you may assume that each player has the same chance of winning each round. Construct a suitable tree diagram and hence determine the probability of each of the following:

   **a**  Laurie wins both rounds
   **b**  Rob wins at least one of the rounds
   **c**  Laurie wins neither given that Steven wins the second round
   **d**  Steven wins the second round given that Laurie wins neither

**4** Envelope I contains 4 letters:  1 A, 1 B, 1 C and 1 D.
Envelope II contains 4 letters:  1 C, 2 Ds and an E.

A two letter 'word' is formed, the first letter being randomly chosen from envelope I and the second letter randomly chosen from envelope II. Construct a suitable tree diagram and hence determine the probability that:

   **a**  one of the two letters is an A
   **b**  the letters chosen are the same
   **c**  the first letter chosen is a D given that the two letters are the same
   **d**  the second letter chosen is a D given the two letters are not the same

**5** The five letters of the word EXACT are written on five cards, with one letter on each card. The five cards are then shuffled and two of the cards are dealt face up in a line to form a 'word'. Construct a suitable tree diagram and hence determine the probability that the 'word' so formed

   **a**  is the word AT
   **b**  starts with an E
   **c**  ends with a T
   **d**  starts with an E and ends with a T
   **e**  starts with an E given that it ends with a T
   **f**  starts with a T given that it ends with an E
   **g**  contains an X given that it contains an A

ISBN 9780170390330

# Use of the words 'not', 'and' and 'or' in probability questions

## Use of the word 'not'

As was mentioned in the *Preliminary work*, prior to chapter one:

If the probability of an event occurring is a then the probability of it **not** occurring is $1 - a$.

Thus the phrase 'not A' in probability is similar to A′, the complement of A, in sets.

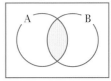
Shading shows A′.

## Use of the word 'and'

If we require the probability that events A **and** B occur we must look at those situations in which *both* A and B occur.

Thus the phrase 'A and B' in probability is similar to A ∩ B, in sets.

Shading shows A ∩ B.

## Use of the word 'or'

Consider the following situations:

- Young Jim notices two bars of chocolate on the kitchen work surface.

  Jim:        '*Mum, can I have a bar of chocolate?*'

  Mum:     '*Yes dear, you can have the Mars Bar **or** the Kit Kat.*'

  Do you think Jim's mum means him to have *both* bars?

- Ms Swift, a deputy principal in XYZ high school, makes the following announcement over the school public address to all year 11 classes:

  Ms Swift    *I would like any students doing year 11 Physics **or** year 11 Art to attend an important meeting in room 3 at the beginning of lunchtime.*'

  Toni Collinge is a student at XYZ and does *both* year 11 Physics and year 11 Art. Do you think Ms Swift expects Toni to attend?

In the first situation Jim's mum would probably not be at all pleased if Jim had both bars. Her use of the word 'or' probably meant 'one or the other but not both'.

In the second situation Ms Swift needed to meet with any student doing Physics or Art. Her use of the word 'or' probably meant 'one or the other or both'.

In these two situations it is only our commonsense understanding of the context and the likely intention behind the use of the word 'or' that allows us to interpret 'or' to mean 'one or the other but not both' in one situation and 'one or the other or both' in the other.

To avoid this possible confusion, when the word 'or' is used in mathematics we interpret 'A or B' to mean 'one or the other or both'. Hence in probability questions the word 'or' should be taken as including the possibility of both i.e. it can be interpreted as meaning 'at least one of'.

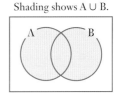
Shading shows A ∪ B.

Thus the phrase 'A or B' in probability is similar to A ∪ B in sets. (Note how the use of ∪ and ∩ in sets avoids the problem.)

Two-way tables

## EXAMPLE 6

The table on the right shows the 36 equally likely number pair outcomes for rolling one red die and one blue die.

Determine the probability that one roll of the two dice will result in:

**a** a 3 on the blue and a 5 on the red

**b** a 3 on the blue or a 5 on the red

**c** neither die showing a 5

| | | | | BLUE DIE | | | |
|---|---|---|---|---|---|---|---|
| | | **1** | **2** | **3** | **4** | **5** | **6** |
| **RED DIE** | **1** | (1, 1) | (1, 2) | (1, 3) | (1, 4) | (1, 5) | (1, 6) |
| | **2** | (2, 1) | (2, 2) | (2, 3) | (2, 4) | (2, 5) | (2, 6) |
| | **3** | (3, 1) | (3, 2) | (3, 3) | (3, 4) | (3, 5) | (3, 6) |
| | **4** | (4, 1) | (4, 2) | (4, 3) | (4, 4) | (4, 5) | (4, 6) |
| | **5** | (5, 1) | (5, 2) | (5, 3) | (5, 4) | (5, 5) | (5, 6) |
| | **6** | (6, 1) | (6, 2) | (6, 3) | (6, 4) | (6, 5) | (6, 6) |

### Solution

**a** One of the 36 number pairs, $(5, 3)$, is a 3 on the blue and a 5 on the red.

$$P(\text{a 3 on the blue and a 5 on the red}) = \frac{1}{36}.$$

**b** 11 of the 36 number pairs involve a 3 on the blue or a 5 on the red.

$$P(\text{a 3 on the blue or a 5 on the red}) = \frac{11}{36}.$$

**c** In 25 of the 36 number pairs neither die shows a 5.

$$P(\text{neither die showing a 5}) = \frac{25}{36}.$$

## Exercise 9D

**1** Use the table from Example 6 to determine the probability that one roll of the two dice will result in:

**a** not getting a 3 on the blue die

**b** a 5 on the red and a 1 on the blue

**c** a 1 on the red or a 5 on the blue

**d** a 4 on the red and a number bigger than 4 on the blue

**e** a 4 on the red or a number bigger than 4 on the blue

**2** The table on the right shows the 12 equally likely outcomes for rolling a fair die and tossing a fair coin.

Determine the probability that when the die is rolled and the coin is tossed the result will be

| | | | | DIE | | | |
|---|---|---|---|---|---|---|---|
| | | **1** | **2** | **3** | **4** | **5** | **6** |
| **COIN** | **Head** | H, 1 | H, 2 | H, 3 | H, 4 | H, 5 | H, 6 |
| | **Tail** | T, 1 | T, 2 | T, 3 | T, 4 | T, 5 | T, 6 |

**a** a head on the coin

**b** a two on the die

**c** a tail on the coin and a 6 on the die

**d** a tail on the coin or a 6 on the die

**e** a head on the coin and an odd number on the die

**f** either a 6 on the die or a head on the coin but not both of these things

ISBN 9780170390330

**3** A normal pack of 52 cards is shuffled and a card is selected at random.

The 52 equally likely outcomes are shown below:

Hearts          A♥, 2♥, 3♥, 4♥, 5♥, 6♥, 7♥, 8♥, 9♥, 10♥, J♥, Q♥, K♥

Diamonds        A♦, 2♦, 3♦, 4♦, 5♦, 6♦, 7♦, 8♦, 9♦, 10♦, J♦, Q♦, K♦

Spades          A♠, 2♠, 3♠, 4♠, 5♠, 6♠, 7♠, 8♠, 9♠, 10♠, J♠, Q♠, K♠

Clubs           A♣, 2♣, 3♣, 4♣, 5♣, 6♣, 7♣, 8♣, 9♣, 10♣, J♣, Q♣, K♣

If one card is selected at random from the shuffled pack, determine the probability that the card is

**a**  the queen of diamonds    **b**  a queen              **c**  a red card

**d**  a three                  **e**  a heart              **f**  a jack

**g**  not a jack               **h**  a jack, queen or king **i**  a red nine

**j**  a red or a nine          **k**  the queen of hearts   **l**  a queen or a heart

**4** The tree diagram on the right shows the eight equally likely outcomes that could result when a fair coin is tossed three times.

Determine the probability that when a coin is tossed three times the outcome is:

**a**  a tail last

**b**  a head first and a tail last

**c**  a head first or a tail last

**d**  the same result on the third toss as was obtained on the second toss

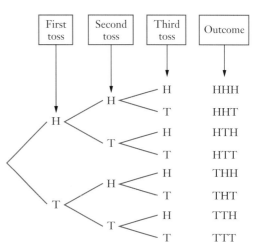

# Tree diagram showing probabilities

Consider the following question:

A bag contains six marbles: 3 red, 2 blue and 1 green. Two marbles are randomly selected from the bag, one after the other, the first marble not being replaced before the second is selected. Determine the following probabilities

**a**  P(red and blue in that order)

**b**  P(two marbles of the same colour)

**c**  P(blue first|blue second)

The tree diagram, though rather large, can be constructed and the required probabilities determined:

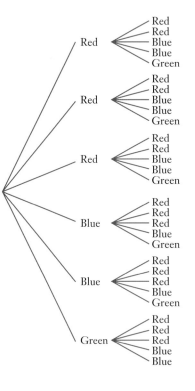

**a**  There are thirty equally likely outcomes.

Six of these thirty are 'red then blue'.

$$P(\text{red and blue in that order}) = \frac{6}{30}$$

$$= \frac{1}{5}$$

**b**  There are thirty equally likely outcomes.

Two marbles of the same colour occur on eight of these.

$$P(\text{two of the same colour}) = \frac{8}{30}$$

$$= \frac{4}{15}$$

**c**  Given the second marble was blue we only need to consider the ten outcomes for which that is the case.

The first marble being blue occurs on two of these ten.

$$P(\text{blue first}|\text{blue second}) = \frac{2}{10}$$

$$= \frac{1}{5}$$

Our tree diagram shows six initial branches, because there are six equally likely outcomes for the first marble out of the bag. A smaller, more manageable tree diagram can be produced by having just three branches to start with, one for each outcome, red, blue, green. We then show the probability of each outcome on the relevant branch of the diagram. A diagram of this form, for the above situation, is shown on the next page.

ISBN 9780170390330

Check carefully that you agree with the probability that is on each branch of the following diagram. Remember the bag initially contains 6 marbles in all, 3 red, 2 blue and 1 green, and the first marble drawn is not being replaced before the second is drawn.

Note the following points:

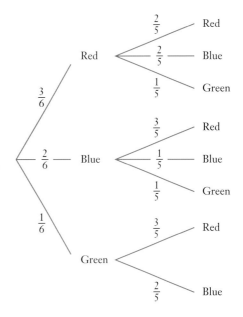

- In a tree diagram it is important that as we move right, the branches leaving each junction are **mutually exclusive** (if one occurs the others cannot) and cover **all** eventualities that can happen from that point (are **exhaustive**). Thus the probabilities on the branches going right from a single point will sum to 1.

  For example, in the tree diagram shown,

  $$\frac{3}{6}+\frac{2}{6}+\frac{1}{6}=1, \quad \frac{2}{5}+\frac{2}{5}+\frac{1}{5}=1, \quad \frac{3}{5}+\frac{1}{5}+\frac{1}{5}=1, \quad \frac{3}{5}+\frac{2}{5}=1.$$

- The probabilities on each branch are assigned with due regard to what has happened up to that point.

  For example, if a red is drawn first the probability of the second marble being red is $\frac{2}{5}$, but if the first is blue then the probability of the second being red is $\frac{3}{5}$.

**The probability of each final outcome is obtained by following the branches to that outcome, and multiplying probabilities along that route:**

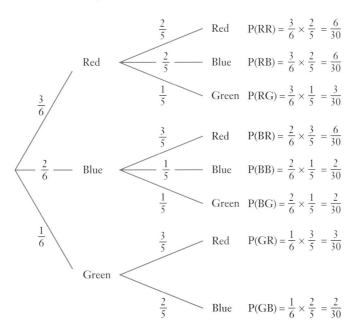

This multiplication of probabilities should seem a reasonable thing to do because each new branch involves a fraction of the probability that applied up to that point. It can be further justified using sets, as shown below.

Consider the Venn diagram on the right.

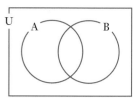

To determine $P(B|A)$ we first consider $P(A)$, because we know event A occurs, and then see how much of $P(A)$ involves B occurring.

i.e.
$$P(B|A) = \frac{P(A \cap B)}{P(A)}$$

Rearranging:
$$P(A \cap B) = P(A) \times P(B|A)$$

This last statement is the *multiplication rule* for probabilities. It tells us that to determine P(both A and B occurring) we multiply the probability of A by the probability of B given A. This is exactly what we do when we multiply probabilities as we pass along the branches of a tree diagram towards a final outcome.

**In the tree diagram we can combine final outcomes by adding probabilities.**

For example: P(both marbles the same colour) $= P(RR) + P(BB) = \dfrac{6}{30} + \dfrac{2}{30} = \dfrac{4}{15}$

Whilst this addition of probabilities should also seem a reasonable thing to do, it can be justified using sets:

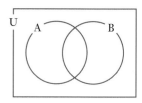

Now
$$n(A \cup B) = n(A) + n(B) - n(A \cap B)$$

∴
$$\frac{n(A \cup B)}{n(U)} = \frac{n(A)}{n(U)} + \frac{n(B)}{n(U)} - \frac{n(A \cap B)}{n(U)}$$

i.e.
$$P(A \cup B) = P(A) + P(B) - P(A \cap B)$$

This last statement is the *addition rule* for probabilities. If A and B are *mutually exclusive events* (cannot occur together) it follows that $P(A \cap B) = 0$ and the rule reduces to:
$$P(A \cup B) = P(A) + P(B)$$

This tells us that to determine P(A or B will occur), where A and B are mutually exclusive events, we add the separate probabilities. In a tree diagram the final outcomes are mutually exclusive so we can combine them in this way.

Note:

- We will see the multiplication rule and the addition rule again later in this chapter. For the moment our tree diagram approach is allowing us to apply the rules intuitively.

- As we would expect, the final probabilities in the tree diagram sum to 1:
$$\frac{6}{30} + \frac{6}{30} + \frac{3}{30} + \frac{6}{30} + \frac{2}{30} + \frac{2}{30} + \frac{3}{30} + \frac{2}{30} = \frac{30}{30} = 1$$

- The probabilities in the tree diagram are sometimes initially best left 'uncancelled' to make combining them easier.

ISBN 9780170390330

- If all we are concerned with in the marbles in the bag situation is the probability of getting (or not getting) reds, the tree diagram could be further simplified as shown on the right.

In this diagram R′ is used to represent the event of the selected marble not being red.

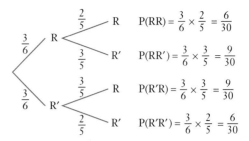

$P(RR) = \frac{3}{6} \times \frac{2}{5} = \frac{6}{30}$

$P(RR') = \frac{3}{6} \times \frac{3}{5} = \frac{9}{30}$

$P(R'R) = \frac{3}{6} \times \frac{3}{5} = \frac{9}{30}$

$P(R'R') = \frac{3}{6} \times \frac{2}{5} = \frac{6}{30}$

The following examples show the use of this idea of a tree diagram with probabilities shown on the branches and the intuitive use of the rules for the multiplication and addition of probabilities.

## EXAMPLE 7

A bag contains ten marbles: 5 red, 3 blue and 2 green. Two marbles are randomly selected from the bag, one after the other, the first marble being replaced before the second is selected. Determine the following probabilities

**a** P(red and blue in that order)

**b** P(red and blue in any order)

**c** P(two marbles of the same colour)

**d** P(two reds | both same colour)

### Solution

First draw a tree diagram:

**a** P(R then B) = 0.15

**b** P(R and B in any order) = P(RB) + P(BR)

$\qquad = 0.15 + 0.15$

$\qquad = 0.3$

**c** P(same colour) = P(RR) + P(BB) + P(GG)

$\qquad = 0.25 + 0.09 + 0.04$

$\qquad = 0.38$

|  | R | P(RR) = 0.25 |
| R | 0.3— B | P(RB) = 0.15 |
|  | G | P(RG) = 0.1 |
|  | R | P(BR) = 0.15 |
| 0.3— B | 0.3— B | P(BB) = 0.09 |
|  | G | P(BG) = 0.06 |
|  | R | P(GR) = 0.1 |
| G | 0.3— B | P(GB) = 0.06 |
|  | G | P(GG) = 0.04 |

**d** Given that both marbles are the same colour we only consider those events. They have probability 0.38 and amongst this 0.38, 0.25 is when two reds occur.

$\therefore$ P(RR | same colour) $= \dfrac{0.25}{0.38}$

$\qquad\qquad\qquad\qquad \approx 0.66$

EXAMPLE 8

The probability event A occurs is 0.5. If event A occurs, the probability of event B occurring is 0.8. If event A does not occur then the probability of event B occurring is 0.3.

These probabilities are shown in the tree diagram on the right with A′ and B′ representing the non occurrence of event A and the non occurrence of event B respectively.

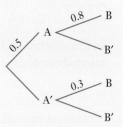

Determine

**a**   $P(A')$

**b**   $P(A \text{ and } B \text{ occurring})$

**c**   $P(B)$

**d**   $P(B'|A)$

**e**   $P(A|B)$

**Solution**

First complete the tree diagram:

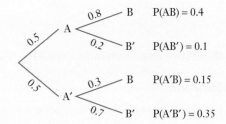

**a**       $P(A') = 0.5$

**b**   $P(A \text{ and } B) = 0.4$

**c**       $P(B) = 0.4 + 0.15$

             $= 0.55$

**d**   $P(B'|A)$ is the 0.2 in the tree diagram.

   $\therefore\ P(B'|A) = 0.2$.

   Alternatively, had we not realised this, the same answer can be obtained from the final column of our tree diagram. Given that A occurs we only consider those events in the final column in which A occurs. These have probability of 0.5 (= 0.4 + 0.1) and amongst this 0.5, 0.1 is for when B′ occurs.

   $\therefore\ P(B'|A) = \dfrac{0.1}{0.5} = 0.2$ as before.

**e**   Given that B occurs we only consider those events in the final column in which B occurs. These have probability of 0.55 (= 0.4 + 0.15) and amongst this 0.55, 0.4 is for when A occurs

   $\therefore\ P(A|B) = \dfrac{0.4}{0.55} = \dfrac{8}{11}$.

EXAMPLE 9

Let us suppose that in a certain breed of chicken, 5% of the chickens are carriers of a particular disease. A test that can accurately detect whether or not a chicken is a carrier of the disease is available but it is expensive to administer.

A cheaper test is developed and trials indicate that this returns a positive result in 90% of the chickens who are carriers of the disease, but unfortunately returns a negative result in the other 10%. When the test is administered to chickens known *not* to be carriers the success rate is 95%. i.e. In 95% of the non carriers a negative result is returned but for the other 5% a positive result is returned.

**a**   A chicken from this breed is selected at random and given this cheaper test. What is the probability the test will return a correct result?

**b**   A chicken from this breed of chickens is selected at random, given this cheaper test, and returns a positive result. What is the probability the chicken is not a carrier of the disease, despite this positive result?

## Solution

The tree diagram is as shown.

Probability

**a**   P(correct result) = 0.045 + 0.902 5

= 0.947 5

**b**   We require: P(not a carrier | +ve result)

Those outcomes that involve a positive result have probability of 0.092 5 (= 0.045 + 0.047 5) and these outcomes contain one with probability 0.047 5 that involves non-carriers.

$$\therefore \text{P(not a carrier} \,|\, \text{+ve result)} = \frac{0.0475}{0.0925}$$

$$\approx 0.514$$

Tree diagram:
- 0.05 → Carrier → 0.9 → +ve result 0.045
- Carrier → 0.1 → −ve result 0.005
- 0.95 → Not a carrier → 0.05 → +ve result 0.0475
- Not a carrier → 0.95 → −ve result 0.9025

You may find this last result interesting. It shows that for the given test, any chicken returning a positive test is still more likely not to be a carrier of the disease than to be a carrier.

## Exercise 9E

**1** The probability event A occurs is 0.4.

If event A occurs, the probability of event B occurring is 0.3.

If event A does not occur then the probability of event B occurring is 0.8.

These probabilities are shown in the tree diagram on the right with A′ and B′ representing the non occurrence of event A and the non occurrence of event B respectively.

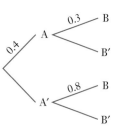

Determine

  **a**   P(A′)

  **b**   P(A and B occurring)

  **c**   P(B)

  **d**   P(B|A)

  **e**   P(A|B)

  **f**   P(A|B′)

**2** The probability event A occurs is 0.6.

If event A occurs, the probability of event B occurring is 0.1.

If event A does not occur then the probability of event B occurring is 0.5.

These probabilities are shown in the tree diagram with A′ and B′ representing the non occurrence of event A and the non occurrence of event B respectively.

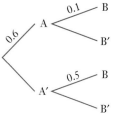

Determine

  **a**   P(A′)

  **b**   P(neither A nor B occurring)

  **c**   P(B)

  **d**   P(at least one of A or B occurring)

  **e**   P(B|A)

  **f**   P(A|B)

**3** A bag contains ten marbles: 7 blue and 3 green. Two marbles are randomly selected from the bag, one after the other, the first marble *not* being replaced before the second is selected.

Determine the probability that the two marbles are

  **a**   blue and green in that order,

  **b**   blue and green in any order,

  **c**   of the same colour,

  **d**   two blues given they are of the same colour.

**4** Bag A contains five marbles: two red and three blue.

Bag B contains five marbles: one red and four blue.

Two coins are tossed. If both coins show heads, a marble is randomly selected from those in bag A, otherwise a marble is randomly selected from bag B. Determine the probability that

  **a**   the selected marble comes from bag B,

  **b**   a blue marble is selected, given the marble selected comes from bag B,

  **c**   the marble selected comes from bag A, given that it was a blue marble.

ISBN 9780170390330

**5** Forty per cent of the students at a particular college are male. Eighty per cent of the males and 40% of the females are studying course A. If one student is randomly selected from the students at this college determine the probability that the student is

**a** female,

**b** studying course A,

**c** male given they are studying course A,

**d** female given they are not studying course A.

**6** Bag A contains 3 marbles:  one red and two blue.

Bag B contains 4 marbles:  one red and three blue.

Bag C contains 5 marbles:  one red and four blue.

A normal fair die is rolled and if the uppermost face shows 1, 2 or 3 a marble is randomly chosen from bag A, if the die shows 4 or 5 on its uppermost face a marble is randomly chosen from bag B and a 6 on the uppermost face of the die sees a marble randomly drawn from bag C.

Determine the probability that the selected marble is

**a** from bag A,

**b** from bag C,

**c** red,

**d** blue,

**e** red and from bag A,

**f** blue or from bag B,

**g** from bag A given that the marble was red,

**h** from bag B given that the marble was blue.

**7** On a particular course, all students who satisfactorily complete all coursework, assignments and tests then sit exam I.

• Those with a mark ≥ 60% in exam I go on to take exam IIA.

• Those with a mark < 60% in exam I then take exam IIB.

• Two thirds of those taking exam I achieve a mark ≥ 60%.

• In exam IIA, one third of those taking it achieve a mark ≥ 70%.

• In exam IIB, one quarter of those taking it achieved a mark < 45%.

Grades are awarded as follows:

• Achieve a mark ≥ 70% in exam IIA:  Grade A.

• Achieve a mark < 70% in exam IIA:  Grade B+.

• Achieve a mark ≥ 45% in exam IIB:  Grade B–.

• Achieve a mark < 45% in exam IIB:  Grade C.

One student is selected at random from all of the students completing this course, to represent the students at a function.

**a** What is the probability that the student selected got a B grade?

**b** Given that the student selected got a B what is the probability that this was in fact a B+?

**8** A college runs diploma courses each lasting 2 years. The students on these courses are classified either as *first year students* or as *second year students* (but not both). The ratio of first year students to second year students is 5 : 4. Fifty five percent of the first year students, and thirty five percent of the second year students, live at home with their parents. If a student is chosen at random from those at this college on these diploma courses, determine the probability that the student is

**a** a second year student,

**b** a second year student or a student living at home with their parents,

**c** a first year student, given they are living at home with their parents.

**9** Five per cent of the people in a particular high risk category are thought to have a particular disease. In an attempt to detect the disease in its early stages a test is developed to identify those who have it.

For those people in the high risk category who do have the disease, the test shows a 98% success rate, i.e. for those who do have the disease the test returns a positive result 98% of the time, and in just 2% the test wrongly returns a negative result (wrongly suggesting that these 2% do not have the disease when in fact they have it).

For those people in the high risk category who do not have the disease the test shows a 96% success rate, i.e. for those who do not have the disease the test returns a negative result 96% of the time, and in just 4% the test wrongly returns a positive result (wrongly suggesting that these 4% do have the disease when in fact they do not have it).

Determine the probability that a person selected at random from those in the high risk category who have the test:

**a** does not have the disease and returns a negative result in the test.

**b** does not have the disease but returns a positive result in the test.

**c** returns an incorrect result in the test.

A person in this high risk category has the test and receives the news that the test gave a positive result.

**d** Determine the probability that this person really does have the disease.
(Give your answer correct to three decimal places.)

**10** A bag contains ten discs, indistinguishable except for their colour. Four of the discs are white and the rest are red. A disc is randomly selected from the bag. If it is white the process stops. If the disc is not white it is not returned to the bag and a second disc is randomly selected from the nine still in the bag. If this second disc is white the process stops. If this second disc is not white it is not returned to the bag and a third disc is randomly selected from the eight still in the bag. The process stops whatever colour this third disc is. Determine the probability that in this process

**a** the 1st disc is not white,

**b** 3 red discs are selected,

**c** exactly 2 discs are selected,

**d** 2 red and 1 white disc are selected,

**e** 2 red and 1 white disc are selected, given more than one disc resulted.

# Imagining the tree diagram

Large tree diagrams can take time to construct so at times we may instead simply imagine following the various branches to an outcome, without actually drawing the tree diagram.

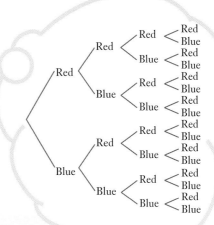

## EXAMPLE 10

A bag contains 10 marbles, 6 red and 4 blue. Four marbles are selected at random, one after the other, with each one selected *not* being returned to the bag before the next is selected.

Find the probability that this will produce

**a**   4 red marbles,          **b**   3 reds then a blue,          **c**   3 reds and 1 blue in any order.

### Solution

'Thinking our way along the appropriate branches':

**a**   $P(R\,R\,R\,R) = \dfrac{6}{10} \times \dfrac{5}{9} \times \dfrac{4}{8} \times \dfrac{3}{7}$

$\qquad\qquad\quad = \dfrac{1}{14} \quad (\approx 0.071\,4)$

**b**   $P(3 \text{ red then blue}) = P(R\,R\,R\,B)$

$\qquad\qquad\qquad\qquad\quad = \dfrac{6}{10} \times \dfrac{5}{9} \times \dfrac{4}{8} \times \dfrac{4}{7}$

$\qquad\qquad\qquad\qquad\quad = \dfrac{2}{21} \quad (\approx 0.095\,2)$

**c**   $P(3 \text{ red and 1 blue}) = P(R\,R\,R\,B) + P(R\,R\,B\,R) + P(R\,B\,R\,R) + P(B\,R\,R\,R)$

$\qquad\qquad\qquad\qquad\quad = \dfrac{2}{21} + \dfrac{6}{10} \times \dfrac{5}{9} \times \dfrac{4}{8} \times \dfrac{4}{7} + \dfrac{6}{10} \times \dfrac{4}{9} \times \dfrac{5}{8} \times \dfrac{4}{7} + \dfrac{4}{10} \times \dfrac{6}{9} \times \dfrac{5}{8} \times \dfrac{4}{7}$

$\qquad\qquad\qquad\qquad\quad = \dfrac{8}{21} \quad (\approx 0.381\,0)$

## EXAMPLE 11

Bag A contains 2 red discs and 3 blue discs.       Bag B contains 1 red disc and 4 blue discs.

A normal die is rolled once. If the outcome is a six a disc is selected from bag A, otherwise a disc is selected from bag B. Find the probability that the selected disc is red.

### Solution

'Thinking our way along the appropriate branches':

$\qquad P(\text{red disc}) = P(6 \text{ on the die then a red disc}) + P(\text{not 6 then red disc})$

$\qquad\qquad\qquad\quad = \dfrac{1}{6} \times \dfrac{2}{5} + \dfrac{5}{6} \times \dfrac{1}{5}$

$\qquad\qquad\qquad\quad = \dfrac{7}{30}$

# Probability rules

Whilst drawing tree diagrams, Venn diagrams, making lists or constructing tables can help our understanding of probability questions and allow us to apply various probability rules intuitively we can apply such rules without the assistance of diagrams and lists if we wish to.

The probability rules are stated below.

### Rule for P(A′)

The *Preliminary work* section reminded us of the rule:

$$P(\text{an event not occurring}) = 1 - P(\text{the event occurring})$$

Using A′ or $\overline{A}$ to represent event A not occurring we can write this as

$$P(A') = 1 - P(A) \qquad \text{or} \qquad P(\overline{A}) = 1 - P(A)$$

### Rule for P(B│A)

As we saw earlier in this chapter, consideration of the Venn diagram on the right leads to the statement:

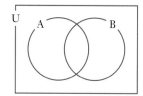

$$P(B\,|\,A) = \frac{P(A \cap B)}{P(A)}$$

### Rule for P(A and B)

Rearranging the above rule leads to the **multiplication rule**:

$$P(A \cap B) = P(A) \times P(B\,|\,A)$$

- Remember that $P(A \cap B)$ means the probability of A **and** B occurring.

- If the probability of B occurring is unaffected by whether or not A has occurred we say that events A and B are **independent**. In such cases $P(B\,|\,A) = P(B)$ and the multiplication rule becomes:

$$P(A \cap B) = P(A) \times P(B)$$

### Rule for P(A or B)

As we saw earlier in this chapter, consideration of the Venn diagram on the right leads to the **addition rule**:

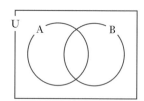

$$P(A \cup B) = P(A) + P(B) - P(A \cap B)$$

- Remember that $P(A \cup B)$ means the probability of A **or** B occurring with the 'or' meaning *at least one of*.

- If A and B are **mutually exclusive**, they cannot occur simultaneously. In such cases we have $P(A \cap B) = 0$ and the addition rule becomes:

$$P(A \cup B) = P(A) + P(B)$$

ISBN 9780170390330

**Summary**

Complementary events. (A and A′):

$$P(A') = 1 - P(A)$$

Conditional probability. (B|A):

$$P(B|A) = \frac{P(A \cap B)}{P(A)}$$

A **and** B (A ∩ B):

To determine the probability of A **and** B occurring we multiply the probabilities together, paying due regard to whether the occurrence of one of the events affects the likelihood of the other occurring:

$$P(A \cap B) = P(A) \times P(B|A)$$

If A and B are **independent** events, $\quad P(B|A) = P(B)$ and so

$$P(A \cap B) = P(A) \times P(B)$$

A **or** B (A ∪ B):

To determine the probability of A **or** B occurring we add the probabilities together and then make the necessary subtraction to compensate for the 'double counting of the overlap':

$$P(A \cup B) = P(A) + P(B) - P(A \cap B)$$

If A and B are **mutually exclusive** events, $P(A \cap B) = 0$ and so

$$P(A \cup B) = P(A) + P(B)$$

These rules can be used to determine probabilities without first drawing tree diagrams, Venn diagrams etc., as the examples which follow will demonstrate. However, a note of caution is appropriate first:

Use the next few pages to gain familiarity with the probability rules but do not be too quick to forsake the various diagrammatic approaches in favour of a purely rules approach. Listing sample spaces, drawing tree diagrams and compiling Venn diagrams may take longer to do but such approaches can greatly clarify a problem, reduce errors and allow you to use the probability rules more intuitively, rather than 'blindly' following a formula.

Shutterstock.com/Arena Photo UK

Alamy Stock Photo/arplus image bank

EXAMPLE 12

The probability of a person having a particular disease is 0.001. If they have the disease the probability they will die from it is 0.7. What is the probability the person has the disease and will die from it?

**Solution**

Suppose event A is 'has the disease' and event B is 'will die from the disease'.

We are given that $\qquad$ P(A) = 0.001 and P(B|A) = 0.7 and we require P(A ∩ B).

Using $\qquad$ P(A ∩ B) = P(A) × P(B|A) = 0.001 × 0.7 = 0.000 7

The probability the person has the disease and will die from it is 0.000 7.

Check that constructing a tree diagram also gives this answer.

EXAMPLE 13

A box contains 50 items, five of which are defective. Two items are randomly chosen from the box, one after the other, the first not being replaced before the second is selected. What is the probability that the two selected will both be defective?

**Solution**

Applying the rule $\qquad$ P(A ∩ B) = P(A) × P(B|A):

P(1st defective and 2nd defective) = P(1st defective) × P(2nd defective | 1st defective)

$$= \frac{5}{50} \times \frac{4}{49}$$

$$= \frac{2}{245} \quad (\approx 0.008\,2)$$

Check that you can also obtain this answer using a tree diagram.

EXAMPLE 14

The probability that in a particular piece of machinery component A will fail is 0.05. Provided A does not fail the probability that component B will fail is 0.01. However if A does fail then the probability of B failing rises to 0.1. What is the probability that

**a** A fails and B does not fail? $\qquad$ **b** A does not fail and B does not fail?

**Solution**

Given: P(A fails) = 0.05, P(B fails | A not fail) = 0.01, P(B fails | A fails) = 0.1.

**a** $\quad$ P(A fails and B not fail)
= P(A fails ∩ B not fail)
= P(A fails) × P(B not fail | A fails)
= 0.05 × 0.9
= 0.045

**b** $\quad$ P(A not fail and B not fail)
= P(A not fail ∩ B not fail)
= P(A not fail) × P(B not fail | A not fail)
= 0.95 × 0.99
= 0.9405

Check that constructing a tree diagram also gives these answers.

## EXAMPLE 15

Events A and B are such that $P(A) = 0.65$, $P(B) = 0.6$ and $P(A \cap B) = 0.4$. Determine

**a** $P(A \cup B)$        **b** $P(\overline{A \cup B})$        **c** $P(A \mid B)$.

**Solution**

**a** $P(A \cup B) = P(A) + P(B) - P(A \cap B)$
$$= 0.65 + 0.6 - 0.4$$
$$= 0.85$$

**b** $P(\overline{A \cup B}) = 1 - P(A \cup B)$
$$= 1 - 0.85$$
$$= 0.15$$

**c** $P(A \mid B) = \dfrac{P(A \cap B)}{P(B)}$
$$= \dfrac{0.4}{0.6}$$
$$= \dfrac{2}{3}$$

Check that using a Venn diagram also gives these answers.

## EXAMPLE 16

A card is randomly drawn from a normal pack.

What is the probability that the card is

**a** a red or a two?        **b** a two or a jack?

**Solution**

**a** P(a red or a two)
= P(red $\cup$ two)
= P(red) + P(two) − P(red $\cap$ two)
$$= \frac{26}{52} + \frac{4}{52} - \frac{2}{52}$$
$$= \frac{7}{13}$$

**b** P(a two or a Jack)
= P(two $\cup$ jack)
= P(two) + P(jack) − P(two $\cap$ jack)
$$= \frac{4}{52} + \frac{4}{52} - \frac{0}{52}$$
$$= \frac{2}{13}$$

Notice that in the last example P(two $\cap$ jack) = 0 because the card we are selecting cannot be both a two and a jack. The events 'card is a two' and 'card is a jack' are **mutually exclusive**.

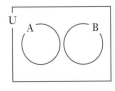

# Independent events

We know that for independent events we obtain the probability of A and B occurring by multiplying the separate probabilities together but how do we know if two events are independent?

In some situations we will intuitively know the events involved are independent. For example, if we roll a normal die and toss a coin, the outcome of the coin tossing is independent of the roll of the die. The probability of getting a head on the coin is 0.5 whatever the result of rolling the die.

Similarly we would expect the result of tossing a coin to be independent of the result of any previous toss. On the other hand, if we were selecting two marbles from a bag containing red and blue marbles, without replacing the first before the second is drawn, we know that probabilities associated with the colour of the second marble depend on the colour of the first marble.

### EXAMPLE 17

If a coin is flipped four times what is the probability of getting four heads?

**Solution**

The result of each flip of the coin is independent of previous flips.

Thus $P(H\ H\ H\ H) = P(H) \times P(H) \times P(H) \times P(H) = \frac{1}{2} \times \frac{1}{2} \times \frac{1}{2} \times \frac{1}{2} = \frac{1}{16}$

If the dependence or independence of two events is not obvious it may be stated in the question, as in the next example.

### EXAMPLE 18

Three students, Alex, Bill and Con, each take their driving test. Their instructor estimates the probability of each of them passing is as follows:

P(Alex passes) = 0.9, P(Bill passes) = 0.8, P(Con passes) = 0.6.

If these events are independent of each other determine the probability that

**a**  Alex and Bill will pass but Con will not

**b**  all three people will pass

**Solution**

**a**  P(Alex and Bill pass and Con fails) $= P(\text{Alex pass}) \times P(\text{Bill pass}) \times P(\text{Con fail})$
$= 0.9 \times 0.8 \times 0.4$
$= 0.288$

**b**  P(All three people pass) $= P(\text{Alex pass}) \times P(\text{Bill pass}) \times P(\text{Con pass})$
$= 0.9 \times 0.8 \times 0.6$
$= 0.432$

### EXAMPLE 19

A manufactured item consists of five parts, A, B, C, D and E. The probability of these parts being defective is 0.01, 0.2, 0.1, 0.02 and 0.01 respectively. The parts are manufactured by different companies so assume the occurrence of defective items are independent of each other. If we randomly select one of each item find the probability that all five are not defective.

**Solution**

P(all okay) = P(A okay) × P(B okay) × P(C okay) × P(D okay) × P(E okay)
$= 0.99 \times 0.8 \times 0.9 \times 0.98 \times 0.99$
$\approx 0.7$

**EXAMPLE 20**

A die is rolled and a coin is tossed. Find the probability of obtaining a six on the die or a head on the coin.

**Solution**

$$P(6 \text{ or } H) = P(6 \cup H)$$
$$= P(6) + P(H) - P(6 \cap H)$$
$$= P(6) + P(H) - P(6) \times P(H) \qquad \text{because } P(6) \text{ and } P(H) \text{ are independent}$$
$$= \frac{1}{6} + \frac{1}{2} - \frac{1}{6} \times \frac{1}{2}$$
$$= \frac{7}{12}$$

The reader should confirm that the same answer can be obtained by creating a table of equally likely outcomes.

## Exercise 9F

Initially attempt the following questions without drawing diagrams.
Instead practise 'imagining the tree diagram' and also 'a rules approach'.

Hint: Remember that if you know P(A) then P(A′) = 1 – P(A).
Use of this rule can sometimes save a lot of time.

**1** A bag contains 10 marbles: 6 red and 4 blue. Two marbles are randomly selected from the bag, the first not being replaced before the second is drawn. Determine the probability of getting

   **a**  two reds,

   **b**  two of the same colour,

   **c**  no blues,

   **d**  at least one blue.

**2** A box contains 100 carburettors, four of which are faulty. A quality control person takes a carburettor from the box, tests it and then puts it to one side. This is repeated until four of the carburettors have been tested.

Correct to four decimal places, what is the probability that

   **a**  none of the four tested are faulty?

   **b**  at least one of the four is faulty?

**3** A box contains 200 carburettors, three of which are faulty. A quality control person takes a carburettor from the box, tests it and then puts it to one side. This is repeated until five of the carburettors have been tested.

Correct to four decimal places, what is the probability that

   **a**  none of the five tested are faulty?

   **b**  at least one of the five is faulty?

**4** A bag contains 10 marbles: 4 red, 3 blue and 3 green. A marble is randomly selected and not replaced, a second is randomly selected and not replaced, and then a third is randomly selected. Determine the probability of getting

   **a**  three reds,

   **b**  three of the same colour,

   **c**  no reds,

   **d**  at least one red.

**5** A bag contains 10 marbles: 4 red, 3 blue and 3 green. Three marbles are randomly selected from the bag, one after the other, with each marble selected being put back into the bag before the next is selected. Determine the probability of getting

   **a**   three reds,                     **b**   three of the same colour,

   **c**   no reds,                       **d**   at least one red.

**6** In a particular school the probability of a randomly selected student being in year 8 is 0.24. If a year 8 student is chosen at random the probability they are male is 0.52.

What is the probability that a randomly selected student from this school is

   **a**   a year 8 male?                     **b**   a year 8 female?

**7** Bag A contains 1 red disc and 3 blue discs. Bag B contains 2 red discs and 2 blue discs.

A normal die is rolled once. If the outcome is a five or a six a disc is selected from bag A, otherwise a disc is selected from bag B. Find the probability that the selected disc is red.

**8** The probability of a randomly selected person having disease X is 0.001.

A test is developed to detect whether or not a person has disease X. For people with the disease the test returns a positive result 98% of the time and wrongly returns a negative result for the other 2%. For people who do not have the disease the test returns a negative result 99% of the time and wrongly returns a positive result for the other 1%.

If a person is chosen at random and given this test what is the probability they will return a positive result?

**9** A fair coin is tossed and a normal die is rolled.

Event A is that of the coin landing head uppermost.
Event B is that of the uppermost face of the die showing a number less than 5.
Determine

   **a**  $P(A)$          **b**  $P(B)$          **c**  $P(A \cap B)$          **d**  $P(A \cup B)$.

**10** Two normal dice, one red and the other blue, are rolled.

Event A is that of the uppermost face of the red die showing a number less than 3.
Event B is that of the uppermost face of the blue die showing an even number.
Determine

   **a**  $P(A)$          **b**  $P(B)$          **c**  $P(A \cap B)$          **d**  $P(A \cup B)$.

**11** Events A and B are such that $P(A) = 0.4$, $P(B) = 0.5$ and $P(A \cap B) = 0.1$.
Determine

   **a**  $P(A \cup B)$          **b**  $P(\overline{A \cup B})$          **c**  $P(A|B)$          **d**  $P(B|A)$.

**12** Events A and B are such that $P(A) = 0.5$, $P(B) = 0.8$ and $P(A \cup B) = 0.9$.
Determine

   **a**  $P(A \cap B)$          **b**  $P(A|B)$          **c**  $P(B|A)$.

**13** The fuel tank of a particular rocket has five seals, A, B, C, D and E that prevent leaks from the fuel tank. The likelihood of any one of these seals failing is independent of the behaviour of the other seals. During a launch these seals are under most strain and the engineers estimate that the probability of each seal failing is:

$P(A) = 0.02$, $P(B) = 0.2$, $P(C) = 0.15$, $P(D) = 0.01$, $P(E) = 0.005$.

Determine the probability that during a launch

**a** all of the seals will fail,

**b** none of the seals will fail (answer rounded to two decimal places),

**c** at least one will fail (answer rounded to two decimal places).

Alamy Stock Photos/Stocktrek Images, Inc.

**14** One child rolls a normal fair die and two other children, independently of each other, write down which number from 1 to 6 they think the die will show on its uppermost face. What is the probability that

**a** both of these children guess correctly?

**b** neither guess correctly?

**c** at least one of them guess correctly?

**15** Two components, X and Y, are manufactured independently of each other. For each type the probability of a randomly chosen component being defective is:

$$P(X \text{ defective}) = 0.005, \quad P(Y \text{ defective}) = 0.01.$$

Determine the probability that for a randomly selected X and a randomly selected Y

**a** both components are defective,

**b** neither of the components are defective,

**c** at least one of the components is defective.

**16** Three components, X, Y and Z, are manufactured independently of each other. For each type the probability of a randomly chosen component being defective is:

$$P(X \text{ defective}) = 0.005, \quad P(Y \text{ defective}) = 0.01, \quad P(Z \text{ defective}) = 0.002.$$

Determine the probability that if we randomly select one of each of these three components

**a** all three components are defective,

**b** none are defective (round to 3 decimal places),

**c** at least one is defective (round to 3 decimal places).

**17** Bags A and B each contain five coloured discs. Bag A contains 3 green and 2 yellow. Bag B contains 1 green and 4 yellow.
A normal die is rolled once and, if the result is even, one disc is randomly selected from bag A.
If the result is odd one disc is randomly selected from bag B.
Determine the probability of this process producing:

**a** a disc from bag B,

**b** a yellow disc from bag B,

**c** a yellow disc,

**d** a yellow disc or a disc from bag B.

**18** Bags A and B each contain five coloured discs. Bag A contains 3 green and 2 yellow. Bag B contains 1 green and 4 yellow.

A normal die is rolled once and, if the result is greater than 4, one disc is randomly selected from bag A. If the result is not greater than 4 one disc is randomly selected from bag B.

Determine the probability of this process producing:

   **a**  a yellow disc from bag B,           **b**  a yellow disc or a disc from bag B.

**19** Events A and B are such that $P(A') = 0.35$, $P(B) = 0.34$ and $P(A \cup B) = 0.86$.
Determine $P(A \cap B)$ and $P(B|A)$.

**20** Events A and B are such that $P(B|A) = 0.20$, $P(A|B) = 0.25$, and $P(A \cap B) = 0.10$.
Determine $P(A \cup B)$.

**21** Events A and B are such that $P(B|A) = \dfrac{1}{4}$, $P(A|B) = \dfrac{2}{5}$ and $P(A \cap B) = \dfrac{3}{22}$.
Determine $P(A \cup B)$.

# More about independent events

As mentioned earlier, for independent events A and B,

$$P(B|A) = P(B)$$

and the rule $\qquad\qquad P(A \cap B) = P(A) \times P(B|A)$

reduces to: $\qquad\qquad P(A \cap B) = P(A) \times P(B).$

It follows that if B is independent of A, then A is independent of B.

i.e. if $\qquad P(B|A) = P(B) \qquad$ then $\qquad\quad P(A|B) = P(A), \qquad$ as proved below:

If $\qquad\quad P(B|A) = P(B) \qquad$ then $\qquad\; P(A \cap B) = P(A) \times P(B).$

But $\qquad\quad P(A|B) = \dfrac{P(A \cap B)}{P(B)}$

and so $\qquad P(A|B) = \dfrac{P(A) \times P(B)}{P(B)}$

$\qquad\qquad\qquad = P(A) \qquad$ as required.

> If A and B are independent events then
>
> - $P(B|A) = P(B)$
> - $P(A|B) = P(A)$
> - $P(A \cap B) = P(A) \times P(B)$

We can use these ideas as a test for independence:

> If we can show that $\qquad P(B|A) = P(B)$
> or that $\qquad\qquad P(A|B) = P(A)$
> or that $\qquad\qquad P(A \cap B) = P(A) \times P(B)$
> then A and B are independent events.

**EXAMPLE 21**

Event A is that of rolling a die and getting an even number.
Event B is that of rolling a die and getting a number less than 5.
Prove that A and B are independent events.

**Solution**

$$P(A) = P(\text{even number}) = \frac{3}{6} = 0.5 \qquad P(A|B) = P(\text{even number}|\text{a number} < 5) = \frac{2}{4} = 0.5$$

Thus $P(A) = P(A|B)$ and hence A and B are independent.

The reader should confirm that for these events it is also the case that

$$P(B) = P(B|A) \text{ and } P(A \cap B) = P(A) \times P(B).$$

# Independence suggested

If, with collected data, we were to find that $P(A) \approx P(A|B)$ this could suggest that events A and B are independent.

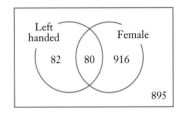

For example, suppose that a survey of one thousand nine hundred and seventy three individuals investigated, amongst other things, left handedness and gender, and gave rise to the Venn diagram shown on the right.

Based on these figures
$$P(\text{Left handed}) = \frac{82 + 80}{1973}$$
$$= 0.082$$

$$P(\text{Left handed}|\text{Female}) = \frac{80}{80 + 916} \qquad\qquad P(\text{Left handed}|\text{Male}) = \frac{82}{82 + 895}$$
$$= 0.080 \qquad\qquad\qquad\qquad\qquad = 0.084$$

The closeness of these figures to each other indicates that whether a person is left handed could well be independent of gender. Approximately 8% of the entire group was left handed and this same percentage was seen within the males and the females.

# More about mutually exclusive events

We have already seen that if events A and B cannot both occur we say they are mutually exclusive. i.e. the occurrence of one of the events excludes the occurrence of the other.

Thus, for mutually exclusive events: $P(A \cap B) = 0$, and the rule $P(A \cup B) = P(A) + P(B) - P(A \cap B)$ reduces to:

$$P(A \cup B) = P(A) + P(B).$$

We can use this as a test for mutual exclusivity:

> If we can show that $\qquad P(A \cap B) = 0$
> or that $\qquad\qquad\qquad P(A \cup B) = P(A) + P(B)$
> then A and B are mutually exclusive events.

If events A and B are such that $P(A) = 0.4$ and $P(B) = P(\overline{A \cup B}) = 0.3$, prove that A and B are mutually exclusive.

**Solution**

If    $P(\overline{A \cup B}) = 0.3$    then    $P(A \cup B) = 0.7$.

But    $P(A) + P(B) = 0.7$.

Thus $P(A \cup B) = P(A) + P(B)$ and hence A and B are mutually exclusive.

Alternatively the given probabilities can be used to complete a Venn diagram and it can be determined that $P(A \cap B) = 0$.

## Exercise 9G

Each of the Venn diagrams below show the probabilities of events A and B occurring. In each case classify events A and B as either independent or dependent.

**1**

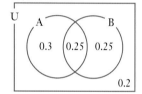

**2**

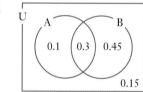

**3**

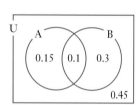

**4**

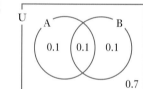

The numbers in the various sections of the following Venn diagrams indicate the probability of the event represented by that section occurring. In each case classify events A and B as either mutually exclusive or not mutually exclusive.

**5**

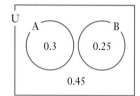

**6**

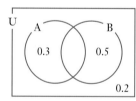

**7**

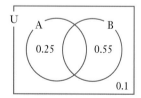

**8**

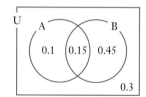

**9** The following pairs of events refer to one roll of a normal die. Which pairs of events are mutually exclusive?

  **a**    obtaining a 3 and obtaining a 4,

  **b**    obtaining an even number and obtaining a number less than 5,

  **c**    obtaining a prime number and obtaining an even number,

  **d**    obtaining a number less than 3 and obtaining a number greater than 5,

  **e**    obtaining a number less than 3 and obtaining a number less than 5.

**10** A bag contains a number of marbles, some red and the rest blue.

Two marbles are randomly selected from the bag, one after the other.

  •    Event A is that of the first marble being red.

  •    Event B is that of the second marble being red.

State whether events A and B are dependent or independent if

  **a**    the first marble is replaced before the second is selected,

  **b**    the first marble is not replaced before the second is selected.

**11** Earlier in this chapter the comment was made that for some situations we intuitively know that two events are independent, for example if we roll a normal die and toss a coin, we know that the outcome of the coin toss is independent of the roll of the die. However, even though the independence is intuitive, use the table of twelve equally likely outcomes shown below to confirm that $P(T) = P(T\,|\,6)$, $P(6) = P(6\,|\,T)$, $P(T \cap 6) = P(T) \times P(6)$

| | | DIE | | | | | |
|---|---|---|---|---|---|---|---|
| | | **1** | **2** | **3** | **4** | **5** | **6** |
| **COIN** | **Head** | H, 1 | H, 2 | H, 3 | H, 4 | H, 5 | H, 6 |
| | **Tail** | T, 1 | T, 2 | T, 3 | T, 4 | T, 5 | T, 6 |

**12** Events A and B are independent events with $P(A) = 0.2$ and $P(B) = 0.25$. Determine

  **a**  $P(A\,|\,B)$     **b**  $P(B\,|\,A)$     **c**  $P(A \cap B)$     **d**  $P(\overline{A \cup B})$

**13** Events A and B are mutually exclusive events with $P(A) = 0.2$, $P(B) = 0.3$. Determine

  **a**  $P(A \cap B)$     **b**  $P(B\,|\,A)$     **c**  $P(A\,|\,B)$     **d**  $P(A \cup B)$

**14** Events A and B are such that $P(A) = 0.25$, $P(B) = 0.5$ and $P(\overline{A \cup B}) = 0.25$.

Prove that A and B are mutually exclusive.

**15** Events A and B are independent events with $P(A) = 0.5$ and $P(B) = 0.6$. Determine

  **a**  $P(A \cap B)$     **b**  $P(A \cup B)$     **c**  $P(B\,|\,A)$     **d**  $P(A\,|\,B)$

**16** Independent events A and B are such that $P(A \cup B) = 0.85$ and $P(A) = 0.25$. Find $P(B)$.

**17** Independent events A and B are such that $P(A \cup B) = 0.4$ and $P(A) = 0.25$. Find $P(B)$.

**18** If $P(A) = 0.2$ and $P(B) = 0.4$ find $P(\overline{A \cup B})$ in each of the following cases:

   **a** A and B are mutually exclusive events,    **b** A and B are independent events.

**19** If $P(A) = 0.2$ and $P(B) = 0.5$ find $P(\overline{A \cup B})$ in each of the following cases:

   **a** A and B are mutually exclusive events,    **b** A and B are independent events.

**20** Let us suppose that for a particular activity the number of equally likely outcomes featuring or not featuring events A and B are as in the table on the right.

Find $x$ if

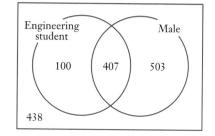

|  | A | A′ |
|---|---|---|
| **B** | $x$ | 5 |
| **B′** | 6 | 2 |

   **a** A and B are mutually exclusive,    **b** A and B are independent.

**21** An analysis of the 1448 students at a college produced the figures shown in the Venn diagram on the right with regards to the numbers of males and females who were, or were not, taking one of the Engineering courses offered by the college (i.e. were or were not classified as an Engineering student).

For a randomly chosen student from this college determine

   **a** P(the student is an Engineering student),

   **b** P(the student is an Engineering student | the student is male),

   **c** P(the student is an Engineering student | the student is female). Comment on your results.

**22** Final year students at a particular college can either follow the *Normal* course in their chosen subject or, if their grades in the previous years have been high enough, they can follow the *Honours* course in that subject. The table below shows the distribution of male and female final year students across these two levels.

|  | Normal course | Honours course | Totals |
|---|---|---|---|
| **Female** | 2814 | 1540 | 4354 |
| **Male** | 1916 | 982 | 2898 |
| **Totals** | 4730 | 2522 | 7252 |

For a randomly chosen final year student from this college determine

   **a** P(the student is on the honours course),

   **b** P(the student is on the honours course | the student is male),

   **c** P(the student is on the honours course | the student is female).

Comment on your results.

**23** The probabilities of events X and Y occurring are as in the Venn diagram: $a + b + c + d = 1$.

Use $a$, $b$, $c$ and $d$ to prove that if $P(X) = P(X|Y)$ then it follows that $P(Y) = P(Y|X)$ and $P(X \cap Y) = P(X) \, P(Y)$.

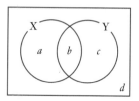

# Miscellaneous exercise nine

This miscellaneous exercise may include questions involving the work of this chapter, the work of any previous chapters, and the ideas mentioned in the Preliminary work section at the beginning of the book.

**1** Events A and B are such that $P(A) = 0.7$, $P(B) = 0.6$ and $P(A \cup B) = 0.8$. Determine

   **a** $P(A \cap B)$    **b** $P(\overline{A \cap B})$    **c** $P(A|B)$    **d** $P(\overline{A}|B)$    **e** $P(A|\overline{B})$

**2** Events A and B are such that $P(A) = 0.45$, $P(B) = 0.2$ and $P(A \cup B) = 0.56$.
Prove that A and B are independent events.

**3** Solve the quadratic equation $2x^2 - x - 36 = 0$ three times: once using the method of completing the square, once using the quadratic formula, and once using factorisation.

**4** A child has four lollies in a bag. Two of the lollies are red, one is green and one is yellow. The child eats the lollies one by one, each time selecting the next one to eat randomly. Determine the probability that the third lolly the child eats is red in each of the following situations.

   **a** The first one the child eats is green and the second one is yellow.

   **b** The first one the child eats is red.

   **c** Nothing is known about the order the lollies are eaten.

**5** State the period and amplitude of each of the following.

   **a**

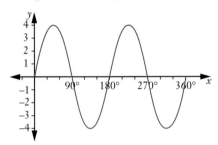

   **b**

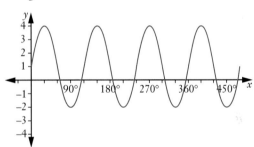

**6** The smallest positive value of $x$ for which $\sin x° = 0.53$ is $x = 32$, to the nearest integer. Without the assistance of your calculator find to the nearest integer all values of $x$ in the interval $-360 \leq x \leq 360$ for which $\sin x° = -0.53$.

**7** Point B(5, −2) is the midpoint of the line joining point A(3, −5) to point C. Find the coordinates of point C.

**8** A company employs 93 people of whom 38 are male. Twenty two of the employees walk to work and 15 of these 22 are female. If one of the 93 employees is chosen at random determine the probability that they are

   **a** female,

   **b** a male who walks to work,

   **c** male given they walk to work,

   **d** someone who walks to work given they are male.

**9** The diagram on the right shows a sketch of
$$y = (x + 2)^2(x - 7).$$

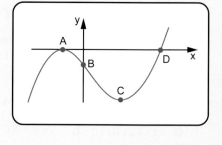

   **a**  Without using a calculator, find the coordinates of point B, the $y$-axis intercept,

   **b**  Without using a calculator, find the coordinates of points A and D, the $x$-axis intercepts.

   **c**  The minimum turning point, C, has coordinates $(a, b)$ where $a$ and $b$ are both integers. Use a graphic calculator to determine $a$ and $b$.

   **d**  Determine the range of values of $p$ for which the equation $(x + 2)^2(x - 7) = p$ has three distinct solutions.

**10** If we assume that when $\theta = \dfrac{\pi}{13}$ radians then $\sin \theta = 0.24$, determine solutions to the equation
$(6 + 25 \sin \theta)(1 - 2 \cos \theta) = 0$ for $0 \le \theta \le 2\pi$.

**11** Solve this question three times: once using a tree diagram approach, once using a Venn diagram approach, once using a rules approach.

In a class of thirty students the teacher is surprised to find that two of the sixteen boys and five of the girls are left handed. (None of the thirty students are ambidextrous.) If one of these thirty students is chosen at random, determine the probability that the chosen student is

   **a**  a left-handed boy,

   **b**  a right-handed girl,

   **c**  left handed given that the chosen student is a girl,

   **d**  a girl given that the chosen student is left handed.

**12** The first stage of a two part random process involves the occurrence, or non occurrence of outcome A. For each of these eventualities the second stage then involves the occurrence, or non occurrence of outcome B. The probabilities associated with some of these events are shown in the tree diagram on the right.

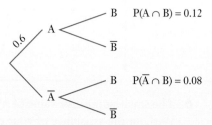

Determine

   **a**  $P(\bar{A} \cap \bar{B})$

   **b**  $P(B)$

   **c**  $P(A \cup B)$

   **d**  $P(B|A)$

   **e**  $P(A|B)$

Are events A and B independent? (Justify your answer.)

ISBN 9780170390330

# 10.

## Counting

- The multiplication principle
- Factorials
- Combinations
- $^nC_r$ and Pascal's triangle
- More about Pascal's triangle
- Miscellaneous exercise ten

Note: Students who are also studying *Mathematics Specialist* will already be familiar with some of the concepts covered in this chapter.

## Situation

A teacher sets her class the challenge of coming up with as many words as they can using some or all of the letters in the word CONSIDER.

Each word has to

- be a proper word that can be found in a dictionary, and

- be of at least 4 letters, and

- have no repeat letters (as the letters in the word CONSIDER each appear only once).
  Thus words like NONE (2Ns), DRESS (2 Ss) and DECIDER (2 Ds and 2Es) are not allowed.

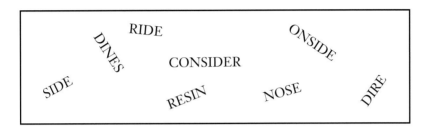

One of the students in the class thought that he would first make a list of **all** the 'words' of four or more letters from the letters in the word CONSIDER, including those that might not be found in a dictionary. Then, using the spell checker on a computer to determine if such a word was a 'real' word, he would cross out any 'illegal' words from the list to end up with his final list.

| | | |
|---|---|---|
| ~~SNOC~~ | ~~CONSI~~ | ONSIDE |
| ~~DISN~~ | ~~RENSI~~ | ~~REDICS~~ |
| NODE | ~~NOCSR~~ | SNORED |
| DONE | ~~DERIS~~ | ~~NSDRCO~~ |
| ~~DCSR~~ | RESIN | ~~NOCDSR~~ |

Try to work out (or at least make some estimate of) how many 'words' would be on the student's list for the spell checker to check.

Hint: Whilst one-, two- and three-letter words are not allowed you might like to consider these situations first in an attempt to establish patterns and techniques that could then be extended to words with four or more letters.

How did you get on with the situation on the previous page? With so many possible 'words' on the student's list it would take a long time if we attempted to count how many there were by first listing them all. Instead we need other ways of determining how many possibilities there are without having to list them all first.

Contrast this to some of the questions in the previous chapter where we did indeed list all of the equally likely outcomes in order to determine probabilities.

Had the situation of the previous page simply involved all possible four-letter words made from a four-letter word, say STEP, we could well have listed all 'words', perhaps in the form of a tree diagram, as shown on the right.

However, with CONSIDER, an eight-letter word, and with four-, five-, six-, seven- and eight-letter words allowed, determining how many 'words' there are by listing would take a long time.

To avoid having to do this we develop techniques for counting the number of possible *arrangements* there are of the letters without having to list them all. Hence the title of this chapter, *Counting*.

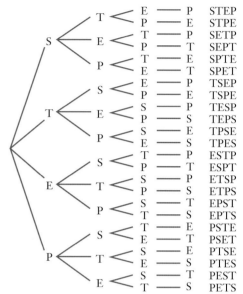

Note in the above tree diagram we have **4** choices of first letter: S, T, E, P.
Having chosen the first letter we then have **3** choices of second letter.
With the first and second chosen we then have **2** choices of third letter.
With first, second and third chosen we have **1** choice for the final letter.

$$\text{Total number of choices} = 4 \times 3 \times 2 \times 1$$
$$= 24.$$

We have obtained the number of possible arrangements using *multiplicative reasoning*. This reasoning is formalised in the *multiplication principle*.

# The multiplication principle

Ordered and unordered selections

> If there are $a$ ways an activity can be performed, and for each of these there are $b$ ways that a second activity can be performed after the first, and for each of these there are $c$ ways that a third activity can be performed after the second, and so on, then there are $a \times b \times c \times \dots$ ways of performing the successive activities.

By appropriately choosing the successive operations we can use this rule to determine the total number of seven-letter 'words' that can be formed using all of the letters of, for example, the word NUMBERS:

The first letter can be chosen in 7 ways, the second can then be chosen in 6 ways, the third in 5 ways etc.

| No. of ways for each letter | | | | | | |
|---|---|---|---|---|---|---|
| 7 | 6 | 5 | 4 | 3 | 2 | 1 |

Total number of words $= 7 \times 6 \times 5 \times 4 \times 3 \times 2 \times 1$
$$= 5040.$$

ISBN 9780170390330

# Factorials

Factorial notation

Use of the multiplication principle frequently involves us in evaluating expressions like:

$$2 \times 1$$

$$3 \times 2 \times 1$$

$$4 \times 3 \times 2 \times 1$$

$$5 \times 4 \times 3 \times 2 \times 1$$

$$6 \times 5 \times 4 \times 3 \times 2 \times 1 \text{ etc.}$$

We write $n!$, pronounced '$n$ **factorial**', to represent

$$n \times (n-1) \times (n-2) \times \ldots \times 3 \times 2 \times 1 \text{ where } n \text{ is a positive integer.}$$

For example

$$3! = 3 \times 2 \times 1$$
$$= 6$$
$$5! = 5 \times 4 \times 3 \times 2 \times 1$$
$$= 120$$
$$10! = 10 \times 9 \times 8 \times 7 \times 6 \times 5 \times 4 \times 3 \times 2 \times 1$$
$$= 3\,628\,800$$

| 3! | |
|---|---:|
| | 6 |
| 5! | |
| | 120 |
| 10! | |
| | 3628800 |

## EXAMPLE 1

Evaluate

**a**  6!

**b**  $5! \div 3!$

**c**  $100! \div 98!$

**Solution**

**a**
$$6! = 6 \times 5 \times 4 \times 3 \times 2 \times 1$$
$$= 720$$

**b**
$$5! \div 3! = \frac{5 \times 4 \times 3!}{3!}$$
$$= 5 \times 4$$
$$= 20$$

**c**
$$100! \div 98! = \frac{100 \times 99 \times 98!}{98!}$$
$$= 100 \times 99$$
$$= 9\,900$$

Now suppose we have 5 objects, all different, that we will call a, b, c, d and e.

We are going to put three of these objects in a row, for example, c e a.

How many different *arrangements* are there of three objects when the three can be chosen from 5 different objects?

Again we could list the possible arrangements:

| | | | | | | | | | |
|---|---|---|---|---|---|---|---|---|---|
| abc | abd | abe | acd | ace | ade | bcd | bce | bde | cde |
| acb | adb | aeb | adc | aec | aed | bdc | bec | bed | ced |
| bac | bad | bae | cad | cae | dae | cbd | cbe | dbe | dce |
| bca | bda | bea | cda | cea | dea | cdb | ceb | deb | dec |
| cab | dab | eab | dac | eac | ead | dbc | ebc | ebd | ecd |
| cba | dba | eba | dca | eca | eda | dcb | ecb | edb | edc |

to arrive at an answer of 60 but again use of the multiplication principle makes the counting process much easier:

Number of arrangements = $5 \times 4 \times 3$

$\qquad\qquad\qquad\qquad = 60$

In this case the number of arrangements is not 5! but instead $\dfrac{5!}{2!}$.

| No. of ways for each letter | | |
|---|---|---|
| 1st | 2nd | 3rd |
| 5 | 4 | 3 |

Thus:

Permutation
calculations

> Whilst the number of arrangements of $n$ different objects is $n!$,
> the number of arrangements of $r$ objects chosen from $n$ different objects is
> $$\frac{n!}{(n-r)!}.$$

For example, the number of arrangements of two different letters that can be made when the two letters can themselves be chosen from the five letters a, b, c, d, e is

$$\frac{5!}{(5-2)!} = \frac{5!}{3!}$$
$$= 5 \times 4$$
$$= 20$$

as we would expect from the multiplication principle.

Note: An arrangement is sometimes referred to as a **permutation**.

## Exercise 10A
Evaluate:

**1** $8!$

**2** $4! \times 2!$

**3** $10! \div 9!$

**4** $10! \div 8!$

**5** $\dfrac{90!}{89!}$

**6** $\dfrac{8!}{6!}$

**7** $3! + 2!$

**8** $\dfrac{100!}{97!}$

**9** $5! - 4!$

ISBN 9780170390330

Express each of the following both numerically and using factorial notation.

**10** The number of five letter arrangements there are of the letters of the word MATHS.

**11** The number of two different letter arrangements there are when the two letters can themselves be chosen from the letters of the word MATHS.

**12** The number of three different letter arrangements there are when the three letters can themselves be chosen from the letters of the word MATHS.

**13** The number of two letter codes there are if the two letters are to be chosen from the 26 letters of the alphabet and the code must involve two different letters.

**14** The number of four letter codes there are if the four letters are to be chosen from the 26 letters of the alphabet and the code must involve four different letters.

**15** The number of permutations there are of the eight letters of the word FORECAST (each letter used once in each permutation).

**16** How many permutations there are, each involving three different digits, if the three digits can themselves be chosen from the set {1, 2, 3, 4, 5, 6, 7, 8, 9}.

# Combinations

Suppose we are given a box of chocolates. We would expect the box to contain a **selection** of chocolates. The way the chocolates were arranged or ordered in the box probably would not concern us too much. If we were to arrange the chocolates differently in the box, or perhaps even empty all of the chocolates into a bag, we would still have the same selection of chocolates.

The arrangement may have changed but the selection is still the same.

iStock.com/kyoshino

Ordered and unordered selections

Combination calculations

A **combination** is a selection — the order does not matter.

A permutation is an arrangement — the order does matter.

Thus, whilst there are 60 possible arrangements, or permutations, of three letters taken from the set {a, b, c, d, e}:

| | | | | | | | | | |
|---|---|---|---|---|---|---|---|---|---|
| abc | abd | abe | acd | ace | ade | bcd | bce | bde | cde |
| acb | adb | aeb | adc | aec | aed | bdc | bec | bed | ced |
| bac | bad | bae | cad | cae | dae | cbd | cbe | dbe | dce |
| bca | bda | bea | cda | cea | dea | cdb | ceb | deb | dec |
| cab | dab | eab | dac | eac | ead | dbc | ebc | ebd | ecd |
| cba | dba | eba | dca | eca | eda | dcb | ecb | edb | edc |

there are just 10 selections, or **combinations**, of three different letters taken from the set {a, b, c, d, e}:

| | | | | | | | | | |
|---|---|---|---|---|---|---|---|---|---|
| abc | abd | abe | acd | ace | ade | bcd | bce | bde | cde |

Notice that to determine the number of **combinations** of three different letters taken from the set {a, b, c, d, e} we had to divide the number of arrangements by 3!, the number of ways of arranging each set of three letters.

Similarly, to determine the number of combinations of $r$ different objects taken from $n$ different objects we divide the number of arrangements by $r!$

But the number of **arrangements** of $r$ different objects taken from $n$ different objects is

$$\frac{n!}{(n-r)!}.$$

Thus the number of **combinations** of $r$ different objects taken from a set containing $n$ different objects will be

$$\frac{n!}{(n-r)!\,r!}.$$

We use the notation $^nC_r$ for the number of combinations of $r$ different objects taken from a set containing $n$ different objects.

There are $^nC_r$ combinations of $r$ objects chosen from $n$ different objects where

$$^nC_r = \frac{n!}{(n-r)!\,r!}.$$

Thus the number of combinations of three objects chosen from five different objects will be

$$^5C_3 = \frac{5!}{(5-3)!\,3!}$$
$$= \frac{5!}{2!\,3!}$$
$$= \frac{5\times4}{2\times1}$$
$$= 10$$

which agrees with our listing on the previous page of the number of combinations of three letters taken from the set {a, b, c, d, e}.

Many calculators can, given the values of $n$ and $r$, determine $^nC_r$.

Get to know how to use your calculator in this regard and use it to confirm the previous answer, $^5C_3 = 10$, and that $^8C_3 = 56$, $^{10}C_4 = 210$, $^{40}C_7 = 18\,643\,560$.

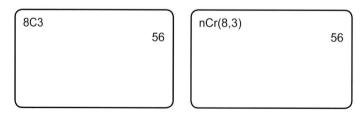

ISBN 9780170390330

## EXAMPLE 2

How many combinations are there of 2 objects chosen from five different objects?

**Solution**

Number of combinations $= {}^5C_2$

$$= \frac{5!}{(5-2)!\ 2!}$$

$$= \frac{5!}{3!\ 2!}$$

$$= 10$$

```
nCr(5,2)
                    10
```

If we label the five objects as A, B, C, D and E, the ten combinations are:

AB,  AC,  AD,  AE,
BC,  BD,  BE,
CD,  CE,
DE.

Note: • ${}^nC_r$ is also written $\binom{n}{r}$. For example $\binom{7}{2} = {}^7C_2$.

• ${}^nC_r$ can be thought of as 'from $n$ **choose** $r$'.

## EXAMPLE 3

A bowl of fruit contains one of each of eight different types of fruit. Parri wants to choose three items of fruit from the bowl to take to school. How many different combinations of three items are possible?

**Solution**

From   8

Choose 3

Number of combinations $= \binom{8}{3}$

$$= \frac{8!}{(8-3)!\ 3!}$$

$$= \frac{8!}{5!\ 3!}$$

$$= 56$$

```
nCr(8,3)
                    56
```

### EXAMPLE 4

From a committee of 20 people a subgroup of 4 is to be formed. How many different subgroups are possible?

**Solution**

From  20

Choose 4

Number of combinations $= {}^{20}C_4$

$$= \frac{20!}{(20-4)!\,4!}$$

$$= \frac{20!}{16!\,4!}$$

$$= 4845$$

```
nCr(20,4)
                    4845
```

## Exercise 10B

**1** How many combinations of four shirts to take on a holiday can be made from the 11 shirts available?

**2** Members of a wine club are invited to select twelve different bottles of wine from a list of 18 wines. How many different selections are possible?

**3** A newspaper editor has 10 pictures available to accompany an article about fishing. He wishes to choose six. How many different selections are possible?

**4** How many selections of 3 chocolates can be made from 15 different chocolates?

**5** From a committee of 12 people a subgroup of 5 is to be formed to represent the committee at a particular function. How many different such subgroups are possible?

**6** For many games of cards a player is dealt a 'hand' of cards from a pack of 52 different cards. The order in which the cards are received is irrelevant, the 'hand' consisting of the cards received, not the order in which they are received. How many different hands of seven cards are there?

**7** A lottery competition involves selecting 6 numbers from 42. The method of selection makes repeat numbers impossible and the order of selection is irrelevant. How many different selections are possible?

**8** Donelle makes a list of 15 people she would like to invite to her party but she is told that she must choose 10. How many different groups of 10 are possible?
Having chosen the ten, and sent out the invitations, two of the chosen say they are unable to attend due to other commitments. She is allowed to choose two replacements from those in the 15 that she initially had to leave off the list.
How many different replacement pairs are there?

ISBN 9780170390330

# $^nC_r$ and Pascal's triangle

Suppose we were asked to expand $(a + b)^5$, that is, to expand: $(a + b)(a + b)(a + b)(a + b)(a + b)$.

We could work through the expansion 'bracket by bracket' or we could determine the coefficients of the various terms and hence complete the expansion using Pascal's triangle, as mentioned in the Preliminary work.

However, even without following either of these approaches, we know that the expansion will be of the form:

$$k_0\, a^5 + k_1\, a^4b + k_2\, a^3b^2 + k_3\, a^2b^3 + k_4\, ab^4 + k_5\, b^5.$$

The first term involves $a^5$ and is obtained by not choosing $b$ from any of the brackets and instead multiplying together the '$a$'s from each bracket. This will occur once in the expansion and so $k_0 = 1$.

The second term involves $a^4b$. Such terms will be obtained when we multiply the $a$ from 4 of the 5 brackets and the $b$ from the other.

We must choose one of the five brackets to supply the $b$. This can be done in $^5C_1$ ways. Thus $k_1 = \,^5C_1$.

The third term involves $a^3b^2$. Such terms will be obtained when we multiply the $a$ from 3 of the 5 brackets and the $b$ from the other 2.

We must choose two of the five brackets to supply $b$. This can be done in $^5C_2$ ways. Thus $k_2 = \,^5C_2$.

Continuing this process leads to:

$$(a + b)^5 = a^5 + \,^5C_1\, a^4b + \,^5C_2\, a^3b^2 + \,^5C_3\, a^2b^3 + \,^5C_4\, ab^4 + \,^5C_5\, b^5$$
$$= a^5 + 5a^4b + 10a^3b^2 + 10a^2b^3 + 5ab^4 + b^5$$

Extending this idea to the general case, $(a + b)^n$, gives the **binomial expansion**:

$$(a + b)^n = a^n + \,^nC_1\, a^{n-1}\, b^1 + \,^nC_2\, a^{n-2}\, b^2 + \,^nC_3\, a^{n-3}\, b^3 + \dots + \,^nC_n\, a^0\, b^n$$

This method does not contradict the Pascal's triangle approach because the numbers in Pascal's triangle could similarly be expressed in $^nC_r$ form as follows:

$$
\begin{array}{ccccccccccc}
&&&&& 1 &&&&& \\
&&&& ^1C_0 && ^1C_1 &&&& \\
&&& ^2C_0 && ^2C_1 && ^2C_2 &&& \\
&& ^3C_0 && ^3C_1 && ^3C_2 && ^3C_3 && \\
& ^4C_0 && ^4C_1 && ^4C_2 && ^4C_3 && ^4C_4 & \\
^5C_0 && ^5C_1 && ^5C_2 && ^5C_3 && ^5C_4 && ^5C_5
\end{array}
$$

## Exercise 10C

Expand

**1** $(a + b)^8$

**2** $(a + b)^{10}$

**3** $(x - y)^8$

**4** $(x + 2y)^6$

**5** $(p - 2q)^6$

**6** $(3x - 2y)^5$

# More about Pascal's triangle

## Polygonal numbers

Placing dots into patterns that form triangles of increasing side length, as shown below, give us the sequence of triangular numbers, 1, 3, 6, 10, 15, …

  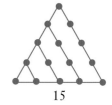

Notice that this sequence of triangular numbers also features in one of the diagonals of Pascal's triangle, as shown on the right.

```
            1
          1   1
        1   2   1
      1   3   3   1
    1   4   6   4   1
  1   5  10  10   5   1
1   6  15  20  15   6   1
```

Now consider the sequence of square numbers:

 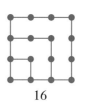 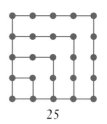

Can you see how the same diagonal of Pascal's triangle, by adding pairs of numbers, can give this sequence of square numbers?

Now consider the pentagonal numbers:

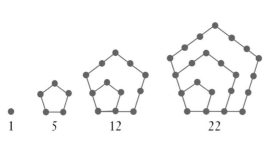 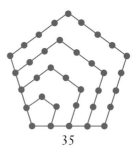

This time use the same diagonal of Pascal's triangle but now double the first number of the pair before adding it to the other.

The sequence of hexagonal numbers, not illustrated here, have the sequence

$$1, 6, 15, 28, 45, 66, …$$

Can you generate this sequence from that same diagonal?

ISBN 9780170390330

## Row totals

If we define the second number in each row of Pascal's triangle as the row number then the row numbers would be as shown below left.

Row no.

| | | | | | | | | | | | | | | | |
|---|---|---|---|---|---|---|---|---|---|---|---|---|---|---|---|
| 0 | | | | | | | 1 | | | | | | = | ?? |
| 1 | | | | | | 1 | + | 1 | | | | | = | ?? |
| 2 | | | | | 1 | + | 2 | + | 1 | | | | = | ?? |
| 3 | | | | 1 | + | 3 | + | 3 | + | 1 | | | = | ?? |
| 4 | | | 1 | + | 4 | + | 6 | + | 4 | + | 1 | | = | ?? |
| 5 | | 1 | + | 5 | + | 10 | + | 10 | + | 5 | + | 1 | = | ?? |

What will be the sum of the numbers in the tenth row?

What will be the sum of the numbers in the twentieth row?

## Another number sequence

Suppose we first align Pascal's triangle somewhat differently:

```
1
1    1
1    2    1
1    3    3    1
1    4    6    4    1
1    5    10   10   5    1
1    6    15   20   15   6    1
1    7    21   35   35   21   7    1
1    8    28   56   70   56   28   8    1
```

And then we 'step' the rows across, as shown below.

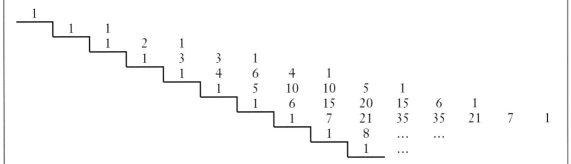

Now find the total for each completed column.

Recognise the sequence?

# Miscellaneous exercise ten

This miscellaneous exercise may include questions involving the work of this chapter, the work of any previous chapters, and the ideas mentioned in the Preliminary work section at the beginning of the book.

**1** Solve each of the following equations.

  **a**   $3x - 2 = 3 - 5x$     **b**   $3(2x + 1) = -5 + 4x$     **c**   $\dfrac{5x - 3}{x - 1} = 4$

  **d**   $\dfrac{2x - 1}{2 - x} = 3$     **e**   $(x - 3)(x + 2) = 0$     **f**   $(x - 1)(x + 5) = 0$

  **g**   $(2x - 1)(x + 7) = 0$    **h**   $(x + 3)(4x - 1)(5x - 9) = 0$   **i**   $x^2 - 6x - 27 = 0$

  **j**   $2x^2 - 3x - 14 = 0$    **k**   $x^3 - 5x^2 - 6x = 0$     **l**   $10x^2 - 7x - 12 = 0$

**2** Determine the rule for each of the straight lines A to H shown in the graph below.

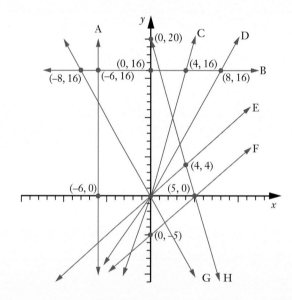

**3** An event has only three possible outcomes, A, B and C, and these outcomes are mutually exclusive. If $P(A) = 2p$, $P(B) = 3p$ and $P(C) = 5p$ determine $p$.

**4** The product of three less than twice a number and seven more than double the number is zero. What could the number be?

**5** From Lookout No.1 a fire is spotted on a bearing 050°. From Lookout No.2 the fire is seen on a bearing 020°. Lookout No.2 is 10 km from Lookout No.1 on a bearing 120°. Assuming that the fire and the two lookouts are all on the same horizontal level find how far the fire is from each lookout.

**6** For a particular experiment three possible outcomes, A, B and C are considered, at least one of these having to be the result. Outcomes A and B can occur together but C is mutually exclusive with A and with B. If $P(A) = \dfrac{1}{2}$, $P(B) = \dfrac{1}{2}$ and $P(A \cap B) = \dfrac{1}{6}$, determine $P(C)$.

**7** Find the equation of the straight line
**a** with a gradient of 3 and cutting the $y$-axis at $(0, 7)$
**b** with a gradient of 3 and passing through the point $(-1, 8)$
**c** passing through $(1, 5)$ and $(3, 1)$
**d** passing through $(4, 8)$ and parallel to $y + 2x = 7$
**e** passing through $(4, 8)$ and perpendicular to $y + 2x = 7$

**8** Solve the quadratic equation $2x^2 + 1 = 4x$ using
**a** completing the square     **b** the quadratic formula
expressing your answers in exact form in each case.

**9** Point $M(5, 7)$ is the midpoint of the straight line AB. If point A has coordinates $(12, 2)$ find the equation of the straight line that is perpendicular to $2x + 3y = 5$ and passes through point B.

**10** Two events X and Y are such that:
$P(X) = 0.6$,   $P(Y) = 0.4$,   $P(\overline{X \cup Y}) = 0.15$,
where $P(X)$ is the probability of event X occurring.
Determine
**a** $P(X \cap Y)$     **b** $P(\overline{Y})$     **c** $P(X | \overline{Y})$

**11** Two normal fair dice are rolled, one red and the other blue, and the two numbers obtained are added together.

- Event A is that of obtaining an even number with the blue die.
- Event B is that of obtaining an even total.
- Event C is that of the total obtained being 12.

For each possible pair of events determine whether the events are independent or not.

**12** Repeat the previous question with events A and B as before but now with event C being that of the total obtained being 7.

**13** If $P(A) = \dfrac{3}{5}$ and $P(B) = \dfrac{1}{3}$, find $P(\overline{A \cup B})$ in each of the following cases:

   **a**  A and B are mutually exclusive events

   **b**  A and B are independent events.

**14** Events A and B are such that $P(A) = 0.6$, $P(B \mid A) = 0.2$ and $P(\overline{A \cup B}) = 0.32$.
Prove that A and B are independent events.

**15** A person is randomly selected from the entire adult population of Australia.
Event A is that of the randomly selected person being male.
Event B is that of the randomly selected person being a professional rugby player.
Determine, with explanation, whether A and B are independent events.

**16 a**  Solve the equation $2p^2 - p - 1 = 0$.

    **b**  Solve the equation $2\cos^2 x - \cos x - 1 = 0$ for $-\pi \le x \le \pi$.

**17** If $f(x) = 2x + 10$ and $g(x) = x^2 - 3x - 4$ determine

   **a**  $f(3)$           **b**  $f(-2)$           **c**  $g(0)$           **d**  $g(3)$

   **e**  $g(-3)$         **f**  $f(2) + g(2)$     **g**  $f(x) + g(x)$    **h**  $f(2x) + g(2x)$

   **i**  the values of $p$ for which $g(p) = 0$       **j**  the values of $q$ for which $f(q) = g(q)$

**18** Given that all of the equations in the 'equations box' are shown graphed below (as unbroken lines) determine the values of $a, b, c, d, e, f, g, h, j, k, m, n, p, q$ and $r$ ($i, l$ and $o$ not used intentionally) of which all but two have integer values.

---

**Equations box**

$y = ax + b$                      $y = cx + 10$                    $y = (x - d)^3$

$y = (x - e)^2 + f$              $y = (x - 1)(x - g)$              $y = \dfrac{h}{x}$

---

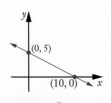
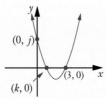

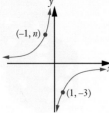
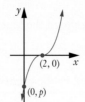
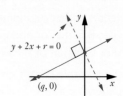

**19** The graph below shows $y = \cos ax$ and $y = \cos [a(x + b)]$ for $0 \le x \le 2\pi$, and with $a$ having the same integer value throughout and $b$ being the smallest possible positive value.

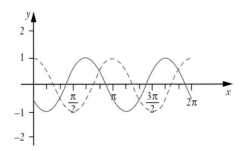

Find $a$ and $b$.

**20** A wheel of radius 60 cm is rotated until a point on the rim that was initially at the lowest point is 80 cm higher than its initial position. Find

**a** the angle in radians through which the wheel is rotated (correct to 2 decimal places)

**b** the length of the circular path travelled by the point (to the nearest cm).

**21** If $\sin (x - y) = \cos x$ prove that $\tan x = \dfrac{1 + \sin y}{\cos y}$.

**22** The display on the right shows the graph of

$$y = \frac{2x + 12}{x - 3}.$$

Without using a graphic calculator, determine the value of $a$, the value of $b$ and the coordinates of points C and D.

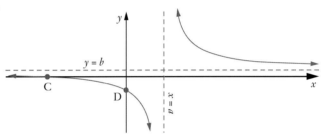

**23** Express the equation of the function shown graphed below in the form $y = a \sin bx + c$.

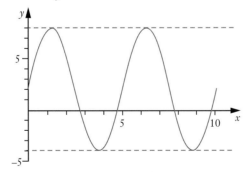

**24** Find $y$, shown in the diagram below, as an exact value.

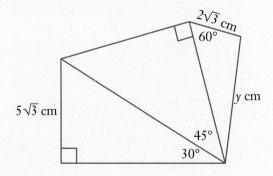

**25** Solve $\sin 2x \cos x + \cos 2x \sin x = 0.5$ for $0 \le x \le \pi$.

**26** Expand and simplify $(x - 2y)^6 + y^2(2x - y)^4$.

**27** (Challenge)

The diagram shows how the vertical motion of a piston can be used to produce rotational motion.

As the piston travels from the low position to the high position and back again the wheel will rotate. If the minor arc PQ is equal in length to $r$, the radius of the wheel, express $x$ as a percentage of $h$ correct to the nearest percent.

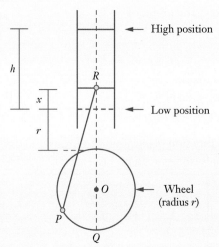

# ANSWERS

Note:

- For questions that do not stipulate a specific level of rounding the answers given here have been rounded to a level considered appropriate for the question.

- If a question asks for an answer to be given 'to the nearest centimetre' it does not necessarily have to be given 'in centimetres' (unless that too is requested). In such a situation an answer of 234.822 centimetres could be written as 235 cm or as 2.35 m, both answers being to the nearest centimetre.

## Exercise 1B    PAGE 11

1 **a** 24°        **b** 49°        **c** 53°
2 **a** 168°       **b** 163°       **c** 147°
3 **a** 30°, 150°  **b** 9°, 171°   **c** 46°, 134°
4 11.2 cm$^2$      5 19.3 cm$^2$    6 18.1 cm$^2$
7 27.7 cm$^2$      8 17.4 cm$^2$    9 138.6 cm$^2$
10 8.7            11 5.9           12 8.0
13 8.6            14 84.9 or 95.1  15 84.7 or 95.3

## Exercise 1C    PAGE 20

1 58             2 12.3           3 54 or 126
4 14             5 75 or 105      6 126
7 6.7            8 75
9 The pole is of length 614 cm, to the nearest centimetre.
10 The two shot journey is 38 metres further than the direct route, to the nearest metre.
11 59            12 14.4          13 43           14 111
15 44            16 62            17 11.9         18 146
19 The boat is then 13.4 km from its initial position, correct to one decimal place.
20 After eight seconds Jim and Toni are 10.7 metres apart, correct to one decimal place.
21 75 or 105     22 99            23 617
24 5.39          25 135           26 80
27 160           28 54
29 The lengths of AC and BC are 672 cm and 824 cm respectively, each answer given to the nearest cm.

30 The smallest angle of the triangle is of size 42°, to the nearest degree.
31 AB is of length 8.1 cm, correct to one decimal place.
32 $a \approx 9.9$ cm, $\angle B \approx 79°$, $\angle C \approx 58°$.
33 Ship B is approximately 15.9 km from ship A.
34 Ship Q is approximately 21.0 km from the lighthouse.
35 To the nearest metre the height of the tower is 21 metres.
36 The parallelogram has diagonals of length 5.1 cm and 9.7 cm, correct to one decimal place.
37 The parallelogram has sides of length 6.8 cm and 10.8 cm, correct to one decimal place.
38 **a** When AC is 2.6 metres $\angle CAB = 20°$, to the nearest degree.
   **b** When AC is 2.1 metres $\angle CAB = 28°$, to the nearest degree.
39 **a** 479 cm   **b** 239 cm   **c** 111 cm   **d** 222 cm
40 **a** At 5 o'clock the distance between the tip of the hour hand and the tip of the minute hand is 155 mm, to the nearest mm.
   **b** At 10 minutes past 5 the distance between the tip of the hour hand and the tip of the minute hand is 119 mm, to the nearest mm.
41 **a** The ship is 1.77 km from the lighthouse, correct to 2 decimal places.
   **b** The ship is 1.17 km from the coastal observation position, correct to 2 decimal places.
42 The largest of the three angles is 98°, to the nearest degree.
43 The height of the tower is 30 metres, to the nearest metre.
44 Point B is 92 metres from point C (nearest metre).
45 $h \approx 20.0$ cm, $\angle H \approx 64°$, $\angle I \approx 61°$ or $h \approx 2.3$ cm, $\angle H \approx 6°$, $\angle I \approx 119°$
46 To the nearest metre B is 141 metres from C.
47 **a** 80°                    **b** 96° (nearest degree)
   **c** 29.2 cm (1 dp)         **d** 52.6 cm$^2$ (1 dp)
48 **a** $x^2 = 244 - 240 \cos \theta$   **b** $x^2 = 277 - 252 \cos \phi$
   **c** 94°
50 The second block has the greater area, by 15 m$^2$ (nearest square metre).

## Exercise 1D   PAGE 28

Answers to numbers **1** to **27** not given here. (You should have checked each one on a calculator.)

**28**  $6\sqrt{3}$ **29**  $2\sqrt{10}$ **30**  $\dfrac{5\sqrt{6}}{2}$

**31**  $2\sqrt{13}$ **32**  $5\sqrt{2}$

## Exercise 1E   PAGE 29

**1 a**  $30°$  **b**  $\dfrac{\sqrt{3}}{3}$

**2 a**  $45°$  **b**  $1$

**3 a**  $60°$  **b**  $\sqrt{3}$

**4 a**  $120°$  **b**  $-\sqrt{3}$

**5 a**  $135°$  **b**  $-1$

**6 a**  $150°$  **b**  $-\dfrac{\sqrt{3}}{3}$

**7**  Gradient of line = tan θ, where θ is the angle or inclination of the line.

## Miscellaneous exercise one   PAGE 30

**1 a**  $11x - 7$ **b**  $x + 23$ **c**  $10x - 3$
 **d**  $13 - 10x$ **e**  $7x + 11$ **f**  $1 - 23x$
 **g**  $x^2 + 8x + 15$ **h**  $x^2 - 2x - 15$ **i**  $2x^2 + 11x + 15$
 **j**  $2x^2 - 11x + 15$

**2 a**  $2(x + 4)$ **b**  $3(2y + 3)$
 **c**  $4a(4b + 3c + 2a)$
 **d**  $a(a + 1)$ **e**  $(x + 8)(x - 1)$ **f**  $(x - 8)(x - 1)$
 **g**  $(x + 7)(x - 2)$ **h**  $(x - 2)(x - 6)$ **i**  $(x + 4)(x - 4)$
 **j**  $2(a + 3)(a - 3)$

**3 a**  $2\sqrt{5}$ **b**  $3\sqrt{5}$ **c**  $10\sqrt{2}$ **d**  $30$
 **e**  $3\sqrt{5}$ **f**  $18\sqrt{2}$ **g**  $21\sqrt{10}$ **h**  $19 + 6\sqrt{2}$

**4**  0.41, 2.35 m

**5**  From ship B, ship A is 9.4 km away on a bearing of 315°.

**6**  No it does not mean that both $C$ and $r$ and $A$ and $r$ are in direct proportion. $C$ and $r$ are in direct proportion because a relationship of the form $C = kr$ for constant $k$ does exist (in this case $k = 2\pi$).
 $A$ and $r$ are not in direct proportion because the rule linking them is not of the form $A = kr$.
 (In this case $A = \pi r^2$ and so $A$ and $r^2$ are in direct proportion.)

**7**  Twelve of the steel frameworks would require a total of 260 metres of steel (to the next 10 metres).

## Exercise 2A   PAGE 36

**1**  10.8 cm **2**  60.3 cm **3**  8.2 cm

**4**  $\dfrac{32\pi}{3}$ cm **5**  $\dfrac{25\pi}{3}$ cm **6**  $\dfrac{28\pi}{3}$ cm

**7**  $24\pi$ cm² **8**  $11\pi$ cm² **9**  $\dfrac{128\pi}{3}$ cm²

**10**  321 cm² **11**  108 cm² **12**  214 cm²
**13**  86 cm² **14**  30 cm² **15**  41 cm²

**16**  $12(2\pi - 3\sqrt{3})$ cm² **17**  $\dfrac{9}{2}(3\pi - 2\sqrt{2})$ cm²

**18**  $\dfrac{25}{3}(5\pi - 3)$ cm²

**19 a**  29.7 cm (1 dp) **b**  65.8 cm (1 dp)

**20**  18.3 cm

**21**  98 cm² to nearest cm² **22**  292 cm² to nearest cm²

**23**  59° **24**  7.3 cm²

**25**  Tip of minute hand travels $12\pi$ cm, tip of hour hand travels $\dfrac{2\pi}{3}$ cm.

**26**  180, 1.85 km **27**  $\dfrac{10\sqrt{5}}{3}$ cm, $6\dfrac{2}{3}$ cm

## Exercise 2B   PAGE 41

**1**  3 rads **2**  1.5 rads **3**  5 rads
**4**  2.5 rads **5**  4 rads **6**  4 rads

**7**  $\dfrac{\pi}{2}$ rads **8**  $\dfrac{\pi}{6}$ rads **9**  $\dfrac{5\pi}{6}$ rads

**10**  $\dfrac{3\pi}{4}$ rads **11**  $\dfrac{\pi}{36}$ rads **12**  $\dfrac{\pi}{10}$ rads

**13**  $\dfrac{4\pi}{9}$ rads **14**  $\dfrac{13\pi}{18}$ rads **15**  45°

**16**  60° **17**  120° **18**  180°
**19**  15° **20**  36° **21**  35°
**22**  70° **23**  0.56 rads **24**  1.10 rads
**25**  2.01 rads **26**  2.97 rads **27**  0.28 rads
**28**  1.47 rads **29**  1.82 rads **30**  0.45 rads
**31**  86° **32**  132° **33**  80°

**34**  34° **35**  $\dfrac{1}{\sqrt{2}}$ **36**  $\dfrac{1}{2}$

**37**  $-\dfrac{1}{\sqrt{2}}$ **38**  1 **39**  $\dfrac{\sqrt{3}}{2}$

**40**  $\dfrac{1}{\sqrt{2}}$ **41**  $\dfrac{1}{\sqrt{2}}$ **42**  $-\sqrt{3}$

**43**  0 **44**  Undefined **45**  $-\dfrac{1}{2}$

ISBN 9780170390330

**46** $-\dfrac{1}{\sqrt{3}}$    **47** $-\dfrac{\sqrt{3}}{2}$    **48** $0$

**49** $\dfrac{1}{2}$    **50** $0$    **51** $0.84$

**52** $-0.42$    **53** $-0.75$    **54** $0.14$

**55** $0.83$    **56** $0.99$    **57** $3.60$

**58** $0.75$    **59** $0.20$ rads    **60** $1.37$ rads

**61** $0.34$ rads    **62** $1.04$ rads

**63 a** $6\pi$ rad/sec   **b** $\dfrac{\pi}{2}$ rad/sec   **c** $\dfrac{\pi}{2}$ rad/sec

**64 a** 1 rev/min   **b** 22.5 rev/min   **c** 10 rev/min

**65** $7.1$    **66** $3.1$    **67** $12.8$

**68** $12.8$    **69** $16.2$    **70** $1.4$

**71 a** $\dfrac{\pi}{2}$ rad   **b** $\dfrac{4\pi}{3}$ rad   **c** $\dfrac{5\pi}{3}$ rad   **d** $\dfrac{11\pi}{6}$ rad

**72 a** $\dfrac{\pi}{4}$ rad   **b** $\dfrac{3\pi}{8}$ rad   **c** $\dfrac{\pi}{20}$ rad   **d** $\dfrac{13\pi}{20}$ rad

**73 a** (Line shown here not to full size.)

$$\overset{B}{\mid}\quad 12 \quad 10 \quad\; 8 \quad\;\; 6 \quad\;\; 4 \quad\;\; 2 \quad\overset{A}{\mid}$$

(The units on the line AB occur every 0.915 cm, starting with zero at A.)

   **b** Yes. Each 1 cm on AB would represent 2 cm diameter, making calibration easier.

## Exercise 2C   PAGE 45

**1** 4 cm      **2** 25 cm

**3** 13.9 cm (1 dp)      **4** 8 cm$^2$

**5** 45 cm$^2$      **6** 114 cm$^2$

**7** 276 cm$^2$ (nearest cm$^2$)      **8** 31.6 cm$^2$ (1 dp)

**9** 39.1 cm$^2$ (1 dp)      **10** 18 cm

**11 a** 90 cm$^2$      **b** 617 cm$^2$

**12 a** 8 cm      **b** 5.1 cm$^2$

**13 a** 6 cm      **b** 3.35 cm$^2$

**14** 80 cm$^2$    **15** 0.37 cm$^2$    **16** 81 cm$^2$

**17** 84 cm$^2$    **18** 26.6 cm$^2$    **19** 16.6 cm$^2$

**20** 14.6 cm$^2$    **21** 11.65 cm$^2$

**22 a** 120 cm      **b** 16 mm

**23** 770 mm$^2$    **24** 16.4 cm    **25** 35%

**26** 269 m$^2$    **27** 233 m$^2$    **28** 125 cm

**29** 16 410 cm$^2$    **30** 177 cm

**31 a** 5 cm      **b** 2 cm

**32** 20.6 cm$^2$    **33** 8.6 %

## Miscellaneous exercise two   PAGE 50

**1 a** $2x^2 + 5x - 3$    **b** $3x^2 + 17x - 28$
   **c** $x^3 + 7x^2 + 7x - 15$    **d** $2x^3 - 9x^2 + 7x + 6$

**2 a** $\dfrac{\sqrt{2}}{2}$   **b** $\dfrac{\sqrt{3}}{3}$   **c** $\dfrac{5\sqrt{2}}{2}$   **d** $2\sqrt{3}$

   **e** $\dfrac{3-\sqrt{5}}{4}$   **f** $\dfrac{3+\sqrt{2}}{7}$   **g** $\dfrac{\sqrt{5}-1}{2}$   **h** $\sqrt{5}-\sqrt{2}$

**3** The topmost point is 35 metres above ground (to the nearest metre).

**4 a** 2.26 m      **b** 1.26 m

**5** Ship B is approximately 30.8 km from ship A, on a bearing N 69° W.

**6** The block has an area of 5270 m$^2$ and a perimeter of 298 metres, both answers given to the nearest integer.

## Exercise 3A   PAGE 56

**1** a, c, e      **2** a, b, e

**3 a** {5, 7, 9, 11}      **b** {8, 10, 12, 14}
   **c** {1}      **d** $\{y \in \mathbb{R}: y \geq 0\}$

**4 a** 18    **b** $-7$    **c** 13
   **d** 4    **e** 21    **f** 23
   **g** $-27$    **h** $5a - 2$    **i** $10a - 2$
   **j** $5a^2 - 2$    **k** 24    **l** $5(a + b) - 2$
   **m** 7    **n** $-2$

**5 a** 9    **b** $-7$    **c** $-3$
   **d** $-3$    **e** 43    **f** 13
   **g** 13    **h** $3(4a - 7)$    **i** $12a - 7$
   **j** $3(a^2 - 12)$    **k** $9a^2 - 12$    **l** $\pm 6$
   **m** 5    **n** $-2$ or 9

**6 a** Function cannot cope with $x < 1$.
   **b** There are no numbers the function cannot cope with.
   **c** Function cannot cope with $x = 0$.
   **d** Function cannot cope with $x = 1$.

**7 a** Function cannot output numbers less than zero.
   **b** Function cannot output numbers less than one.
   **c** Function cannot output zero.
   **d** Function cannot output zero.

**8** $\{y \in \mathbb{R}: 5 \leq y \leq 8\}$      **9** $\{y \in \mathbb{R}: -3 \leq y \leq 0\}$

**10** $\{y \in \mathbb{R}: -6 \leq y \leq 15\}$      **11** $\{y \in \mathbb{R}: 20 \leq y \leq 40\}$

**12** $\{y \in \mathbb{R}: -1 \leq y \leq 9\}$      **13** $\{y \in \mathbb{R}: -4 \leq y \leq 1\}$

**14** $\{y \in \mathbb{R}: 0 \leq y \leq 9\}$      **15** $\{y \in \mathbb{R}: 0 \leq y \leq 16\}$

**16** $\{y \in \mathbb{R}: 1 \leq y \leq 10\}$      **17** $\{y \in \mathbb{R}: 0.25 \leq y \leq 1\}$

**18** $\{y \in \mathbb{R}: y \geq 1\}$      **19** $\{y \in \mathbb{R}: y \geq -1\}$

**20** $\{y \in \mathbb{R}: y \geq 4\}$   **21** $\{y \in \mathbb{R}: y \neq 0\}$

**22** $\{y \in \mathbb{R}: y \neq 1\}$   **23** one-to-one

**24** one-to-one   **25** many-to-one

**26** many-to-one   **27** one-to-one

**28** one-to-one   **29** Domain: $\mathbb{R}$, Range: $\mathbb{R}$

**30** Domain: $\mathbb{R}$, Range: $\{y \in \mathbb{R}: y \geq 0\}$

**31** Domain:$\{x \in \mathbb{R}: x \geq 0\}$, Range:$\{y \in \mathbb{R}: y \geq 0\}$

**32** Domain:$\{x \in \mathbb{R}: x \geq 3\}$, Range:$\{y \in \mathbb{R}: y \geq 0\}$

**33** Domain:$\{x \in \mathbb{R}: x \geq -3\}$, Range:$\{y \in \mathbb{R}: y \geq 0\}$

**34** Domain:$\{x \in \mathbb{R}: x \geq 3\}$, Range:$\{y \in \mathbb{R}: y \geq 5\}$

**35** Domain:$\{x \in \mathbb{R}: x \neq 3\}$, Range:$\{y \in \mathbb{R}: y \neq 0\}$

**36** Domain:$\{x \in \mathbb{R}: x > 3\}$, Range:$\{y \in \mathbb{R}: y > 0\}$

## Miscellaneous exercise three   PAGE 59

**1 a** $x = 11$   **b** $x = -5$

**2** $\{1, -1, -3, -5\}$

**3** For the domain $-2 \leq x \leq 3$ the range is $-1 \leq y \leq 4$.
For the domain $\{-2, -1, 0, 1, 2, 3\}$ the range is $\{-1, 0, 1, 2, 3, 4\}$.

**4 a** $a^2 + 2ab + b^2$   **b** $a^3 + 3a^2b + 3ab^2 + b^3$
  **c** $a^3 + 6a^2b + 12ab^2 + 8b^3$   **d** $a^3 - 6a^2b + 12ab^2 - 8b^3$

**5 a** A function. One-to-one.

  **b** A function. Many-to-one.

  **c** Not a function.

  **d** A function. Many-to-one.

  **e** A function. One-to-one.

  **f** Not a function.

**6** That part of triangle ABC not lying in any of the circles has an area of 4.3 cm$^2$ (correct to the nearest 0.1 cm$^2$).

**7** Ship B is approximately 7.3 km from C on a bearing of 064°.

**8** The block has an area of 6399 m$^2$, to the nearest square metre.

**9** 240 litres

## Exercise 4A   PAGE 68

**1 A: a** $(0, 1)$   **b** 1   **c** $y = x + 1$

  **B: a** $(0, -1)$   **b** 2   **c** $y = 2x - 1$

  **C: a** $(0, 0)$   **b** 0.5   **c** $y = 0.5x$

  **D: a** $(0, 0)$   **b** $-1$   **c** $y = -x$

  **E: a** $(0, 6)$   **b** 3   **c** $y = 3x + 6$

  **F: a** $(0, 2)$   **b** 0   **c** $y = 2$

  **G: a** $(0, -3)$   **b** 1   **c** $y = x - 3$

  **H: a** $(0, -3)$   **b** $-2$   **c** $y = -2x - 3$

  **I: a** $(0, 4)$   **b** 0   **c** $y = 4$

  **J: a** $(0, -3)$   **b** $-0.5$   **c** $y = -0.5x - 3$

  **K: a** $(0, -0.5)$   **b** 1.5   **c** $y = 1.5x - 0.5$

  **L: a** $(0, \frac{4}{3})$   **b** $\frac{1}{3}$   **c** $y = \frac{1}{3}x + \frac{4}{3}$

**2 a** Points lie in a straight line.
    Equation of line is $y = 2x + 5$.

  **b** Points lie in a straight line.
    Equation of line is $y = 5x - 7$.

  **c** Points do not lie in a straight line.

  **d** Points lie in a straight line.
    Equation of line is $y = x - 4$.

  **e** Points lie in a straight line.
    Equation of line is $y = -2x + 10$.

  **f** Points lie in a straight line.
    Equation of line is $y = 5$.

  **g** Points do not lie in a straight line.

  **h** Points lie in a straight line.
    Equation of line is $y = 5x - 13$.

**3**

| Equation | Gradient | y-axis intercept |
|---|---|---|
| $y = 2x + 3$ | 2 | $(0, 3)$ |
| $y = 3x + 4$ | 3 | $(0, 4)$ |
| $y = -2x - 7$ | $-2$ | $(0, -7)$ |
| $y = 6x + 3$ | 6 | $(0, 3)$ |

**4** $y = 4x + 6$   **5** $y = -x - 5$

**6** Lines B, D, E, F and G are in the family, the others are not.

**7** Lines A, D, E, G and H are in the family, the others are not.

**8** $y = -4x - 3$. Yes   **9** $y = 2x - 3$. A, C, D

**10**

| Equation | Written as $y = mx + c$ | Gradient | y-axis intercept |
|---|---|---|---|
| $2y = 4x - 5$ | $y = 2x - 2.5$ | 2 | $(0, -2.5)$ |
| $4y = 3x + 7$ | $y = 0.75x + 1.75$ | 0.75 | $(0, 1.75)$ |
| $3y - 2x = 6$ | $y = \frac{2}{3}x + 2$ | $\frac{2}{3}$ | $(0, 2)$ |
| $4x + 3y - 6 = 0$ | $y = -\frac{4}{3}x + 2$ | $-\frac{4}{3}$ | $(0, 2)$ |
| $3x + 5y = 8$ | $y = -0.6x + 1.6$ | $-0.6$ | $(0, 1.6)$ |

**11** $a = 26, b = 40, c = -2$

**12** $d = 0.5, e = -1, f = -6, g = 1.5, h = 1, i = -5$

**13 a** $P$ and $t$ are directly proportional. The rule is $P = t$.

**b** $P$ and $t$ are not directly proportional.

**c** $P$ and $t$ are directly proportional. The rule is $P = 4t$.

**d** $P$ and $t$ are not directly proportional.

**e** $P$ and $t$ are directly proportional.
The rule is $P = 0.25t$.

**f** $P$ and $t$ are directly proportional.
The rule is $P = 0.75t$.

**g** $P$ and $t$ are directly proportional. The rule is $P = 0.5t$.

**h** $P$ and $t$ are not directly proportional.

## Exercise 4B   PAGE 73

**1 a** $(7, 9)$ **b** $(5, 10)$ **c** $(3, 5)$

**d** $(-2, 1)$ **e** $(-2, 3.5)$ **f** $(12, 1)$

**g** $(8, -5.5)$ **h** $(0, 7.5)$ **i** $(1, 1)$

**2 a** $2$ **b** $-4$ **c** $2$

**d** $0.5$ **e** $-0.25$ **f** $-1$

**g** $-2$ **h** $2.5$ **i** $0.5$

**3 a** 5 units **b** 5 units **c** 13 units

**d** 25 units **e** 17 units **f** 10 units

**g** $5\sqrt{2}$ units ($\approx 7.07$ units)

**h** $\sqrt{58}$ units ($\approx 7.62$ units)

**i** $\sqrt{61}$ units ($\approx 7.81$ units)

**4 a** $2$ **b** $\sqrt{5}$ units ($\approx 2.24$ units) **c** $(3.5, 7)$

**5 a** $1.6$ **b** $\sqrt{89}$ units ($\approx 9.43$ units) **c** $(1.5, 5)$

**6** $-4$ or $12$

**7 a** $\sqrt{82}$ km ($\approx 9.06$ km)

**b** $7\sqrt{2}$ km ($\approx 9.90$ km)

**c** $2\sqrt{10}$ km ($\approx 6.32$ km)

**8** Stage 1 gradient is 0.2, stage 2 gradient is $\frac{5}{9}$, stage 3 gradient is 2.5.

## Exercise 4C   PAGE 76

**1** A: $y = -3$, B: $y = 1$, C: $y = -0.5x + 5$, D: $x = 5$, E: $y = x + 3$, F: $y = 9$, G: $x = -3$, H: $y = 3x + 2$, I: $x = 7$, J: $y = x$

**2** $y = 0$ **3** $x = 0$

**4** $y = 3x + 4$, Yes **5** $y = 0.5x + 2$, D and E

**6 a** $y = x + 2$ **b** $y = -x + 5$

**c** $y = -2x + 8$ **d** $y = 5x + 8$

**e** $y = 0.5x + 5$ **f** $y = -0.5x - 1.5$

**g** $y = 1.5x - 11.5$ **h** $y = -\frac{1}{3}x + \frac{4}{3}$

**7 a** $y = x + 3$ **b** $y = -4x - 1$

**c** $y = -3x + 43$ **d** $y = 2x - 1$

**e** $y = \frac{1}{3}x + \frac{5}{3}$ **f** $y = -2x + 4$

**g** $y = \frac{5}{3}x + 4$ **h** $y = -5x + 5$

**8** $y = 2x - 1$, B and E

**9** $y = 0.5x + 2.5$, $f = 7$, $g = -2$, $h = 13$, $i = -2$, $j = 4.4$.

**10** $(4, 0)$, $y = -2x + 8$

**11** $(6, 0)$, $y = 4x - 24$

**12** $F = 1.8C + 32$

**a** $131°F$ **b** $257°F$ **c** $14°F$

**d** $15°C$ **e** $30°C$ **f** $-40°C$

**13** $A = 0.24N + 40$

**14 a** A($-80, 20$), B($120, 120$), C($-100, 60$), D($-60, -20$), E($100, 160$), F($140, 80$)

**b** $\sim224$ m **c** $y = 0.5x + 60$

**d** $y = -2x - 140$ **e** $y = -2x + 360$

**15** When $t = 2$, $A = 3970$. When $A = 3850$, $t = 10$. $A = -15t + 4000$

**16** $C = 120T + 85$

**17** $P = 4.5N - 3650$

**a** $\$3100$ **b** $\$6925$ **c** $812$

**18 a** $110, 540$ **b** $\$1660$

**19** $k = 0.2$, $L_0 = 0.45$, 5 cm.

## Exercise 4D   PAGE 81

**1** A and E, B and J, C and H, F and K, G and I.

**2** $y = 2x - 5$

**3** A and D, B and G, C and E, F and K, I and J.

**4** $y = -\frac{1}{2}x + 5$

**5** $y = 3x + 5$

**6 a** Point B has coordinates $(2, -1)$.

**b** The required equation is $y = 2x - 5$.

## Miscellaneous exercise four   PAGE 82

**1** A, C, E, F, H, I, J, L

**2** A does not, B does not, C does, D does not, E does.

**3** F does, G does, H does not, I does not, J does.

**4 a** $11$ **b** $-1$ **c** $23$ **d** $-8$

**e** $-28$ **f** $14.5$ **g** $12$ **h** $-21$

**i** $7m - 15$ **j** $m = 6$ **k** $p = 5$ **l** $q = 7$

**m** $r = -3$ **n** $s = 4.5$

**5 a** $(3, -5)$ **b** $(-1, 4)$

**6 a** Domain: $\mathbb{R}$, Range: $\mathbb{R}$

**b** Domain: $\{x \in \mathbb{R}: x \geq 5\}$, Range: $\{y \in \mathbb{R}: y \geq 0\}$

**c** Domain: $\mathbb{R}$, Range: $\{y \in \mathbb{R}: y \geq 0\}$

**d** Domain: $\{x \in \mathbb{R}: x \neq 5\}$, Range: $\{y \in \mathbb{R}: y \neq 0\}$

**e** Domain: $\{x \in \mathbb{R}: x \neq 5\}$, Range: $\{y \in \mathbb{R}: y > 0\}$

**f** Domain: $\{x \in \mathbb{R}: x > 5\}$, Range: $\{y \in \mathbb{R}: y > 0\}$

**8** $a = -1, b = 4, c = 9, d = 19, e = 29, f = 11, g = 99.$

**9** $\dfrac{25}{2}(2\sqrt{3} - \pi)$ cm$^2$

## Exercise 5B   PAGE 90

**1** A: $y = x^2 + 1$, B: $y = x^2 - 2$, C: $y = x^2 - 4$,
D: $y = (x - 3)^2 + 1$, E: $y = (x + 3)^2 - 4$, F: $y = (x - 2)^2 - 3$

**2** G: $y = -x^2$, H: $y = -x^2 + 3$, I: $y = -(x - 3)^2$,
J: $y = -(x + 3)^2 + 1$

**3** K: $y = 2x^2 - 2$, L: $y = 2(x - 3)^2$, M: $y = 2(x + 2)^2$,
N: $y = 2(x - 3)^2 - 2$

**4 a** $y = 3(x + 1)^2 - 4$ **b** $y = -2(x - 3)^2 + 8$

**c** $y = \dfrac{1}{2}(x - 4)^2 - 3$ **d** $y = -\dfrac{1}{2}(x + 2)^2 + 10$

## Exercise 5C   PAGE 97

For questions **1** to **10** the sketches, not shown here, should be consistent with the information obtained in earlier parts of the question.

**1 a** $x = -1$ **b** min at $(-1, -4)$ **c** $(0, -3)$

**2 a** $x = 3$ **b** min at $(3, 5)$ **c** $(0, 14)$

**3 a** $x = 1$ **b** max at $(1, 3)$ **c** $(0, 1)$

**4 a** $(0, 21)$ **b** $(3, 0)$ and $(7, 0)$

**c** $x = 5$ **d** min at $(5, -4)$

**5 a** $(0, -12)$ **b** $(-4, 0)$ and $(3, 0)$

**c** $x = -0.5$ **d** min at $(-0.5, -12.25)$

**6 a** $(0, 8)$ **b** $(-2, 0)$ and $(-4, 0)$

**c** $x = -3$ **d** min at $(-3, -1)$

**7 a** $x = -2$ **b** min at $(-2, -16)$ **c** $(0, -12)$

**8 a** $x = 3$ **b** min at $(3, -8)$ **c** $(0, 1)$

**9 a** $x = 1$ **b** max at $(1, 3)$ **c** $(0, 1)$

**10 a** $x = 2$ **b** max at $(2, 5)$ **c** $(0, -3)$

**11 a,b** Check your answers with those of others in your class and with your teacher.

**c** Check your sketch with a graphic calculator display of the function.

**d** The greatest rectangular area is 49 m$^2$, dimensions 7 m by 7m (i.e. a square).

**12 a,b** Check your answers with those of others in your class and with your teacher.

**c** Check your sketch with a graphic calculator display of the function.

**d** The greatest rectangular area is 50 m$^2$, dimensions are 5 m by 10 m ($x = 5, y = 10$).

**13 a** $(2.5, 11.25)$ **b** 10

**c** Concave down

**14 a** \$590 000 **b** \$545 000

**c** $t = 10$, \$530 000

**15** The maximum value of $h$ is 122.5 and it occurs when $t = 5$.

**16 a** Concave up

**b** The bridge is 15 m above water level.

**c** $x = 40$

**d** From D to C is 40 metres.

**e** From D to E is 80 metres.

**f** From D to A is 30 metres.

**17 a** Concave down

**b** At the midpoint of the bridge $x = 150$

**c** The vertical strut one quarter of the way along the bridge is 10 m long.

**d** Maximum clearance is

**i** 54 m at low tide **ii** 46m at high tide

## Exercise 5D   PAGE 102

**1** Quadratic. $y = x^2 + 6x + 5$

**2** Neither.

**3** Quadratic. $y = x^2 + x + 3$

**4** Linear. $y = 5x + 1$

**5** Quadratic. $y = x^2 + 2$

**6** Linear. $y = \pi x + \pi$

**7** Neither.

**8** Quadratic. $y = x^2 + 5x + 4$

**9** Linear. $y = 8x + 3$

**10** Quadratic. $y = 2x^2 + 3$

**11** Quadratic. $y = 3(x - 2)^2 + 1$

**12** Quadratic. $y = -(x - 3)^2 + 5$

**13 a**

| Length of side of cube ($L$ units) | 1 | 2 | 3 | 4 | 5 | 6 |
|---|---|---|---|---|---|---|
| Surface area of cube ($n$ units$^2$) | 6 | 24 | 54 | 96 | 150 | 216 |

**b** Quadratic **c** $n = 6L^2$

**14 a**

| Number of rows of cans ($r$) | 1 | 2 | 3 | 4 | 5 | 6 |
|---|---|---|---|---|---|---|
| Number of cans ($n$) | 1 | 3 | 6 | 10 | 15 | 21 |

**b** Quadratic      **c** $n = 0.5r^2 + 0.5r$

## Exercise 5E   PAGE 107

**1** $y = (x + 2)^2 - 5$, min $(-2, -5)$
**2** $y = (x - 3)^2 - 7$, min $(3, -7)$
**3** $y = (x - 4)^2 - 6$, min $(4, -6)$
**4** $y = (x + 3)^2 - 6$, min $(-3, -6)$
**5** $y = (x - 1.5)^2 - 0.25$, min $(1.5, -0.25)$
**6** $y = (x - 2.5)^2 - 3.25$, min $(2.5, -3.25)$
**7** $y = -(x - 5)^2 + 24$, max $(5, 24)$
**8** $y = 2(x - 3)^2 - 15$, min $(3, -15)$
**9** $y = -2(x - 2)^2 + 12$, max $(2, 12)$
**10** $y = 2(x + 1.25)^2 + 0.875$, min $(-1.25, 0.875)$

## Miscellaneous exercise five   PAGE 108

**1 a** 31      **b** 1      **c** 44
**2 a** Concave down
  **b** Concave up
  **c** Concave down
**3** $a = 1, b = -1, c = -13, d = 0, e = 9, f = 0$.
**4 a** $-\dfrac{1}{2}$      **b** $-\dfrac{1}{3}$
  **c** 5      **d** $y = 2x + 7$
**5 a** $(0, 3)$      **b** $(1, 0), (3, 0)$
  **c** $x = 2$      **d** min at $(2, -1)$

**6**

| Equation | Cuts $y$-axis | Line of symmetry | Turning point | |
|---|---|---|---|---|
| | | | Coordinates | Max or min? |
| $y = x^2 + 4x + 1$ | $(0, 1)$ | $x = -2$ | $(-2, -3)$ | min |
| $y = x^2 - 2x - 1$ | $(0, -1)$ | $x = 1$ | $(1, -2)$ | min |
| $y = 2x^2 + 4x - 3$ | $(0, -3)$ | $x = -1$ | $(-1, -5)$ | min |
| $y = 2x^2 + 6x - 1$ | $(0, -1)$ | $x = -1.5$ | $(-1.5, -5.5)$ | min |

**7 a** $x = -3$      **b** $(-3, -4)$
  **c** $x = -1$      **d** $(-1, -1)$
**8** A: $x = 4$, B: $y = -3$, C: $y = x$, D: $y = x + 2$, E: $y = 2x + 4$,
F: $y = -x$, G: $y = 0.25x + 4$, H: $y = 0.5x + 1$, I: $y = -0.5x - 1$
**9** I: $y = (x - 1)(x - 3)$, II: $y = (x + 2)(2 - x)$,
III: $y = -(x + 1)(x + 3)$, IV: $y = (x + 1)(x + 3)$

**10 a** Rule: $y = 3x + 4$

| $x$ | 1 | 2 | 3 | 4 | 5 | 6 | 7 | 8 |
|-----|---|---|---|---|---|---|---|---|
| $y$ | 7 | 10 | 13 | 16 | 19 | 22 | 25 | 28 |

**b** Rule: $y = 2x - 1$

| $x$ | 1 | 2 | 3 | 4 | 5 | 6 | 7 | 8 |
|-----|---|---|---|---|---|---|---|---|
| $y$ | 1 | 3 | 5 | 7 | 9 | 11 | 13 | 15 |

**c** Rule: $y = -2x + 17$

| $x$ | 1 | 2 | 3 | 4 | 5 | 6 | 7 | 8 |
|-----|----|----|----|---|---|---|---|---|
| $y$ | 15 | 13 | 11 | 9 | 7 | 5 | 3 | 1 |

**d** Rule: $y = 5x - 1$

| $x$ | 1 | 2 | 3 | 4 | 5 | 6 | 7 | 8 |
|-----|---|---|----|----|----|----|----|----|
| $y$ | 4 | 9 | 14 | 19 | 24 | 29 | 34 | 39 |

**e** Rule: $y = 3x - 2$

| $x$ | 3 | 8 | 1 | 6 | 7 | 4 | 5 | 2 |
|-----|---|----|---|----|----|----|----|---|
| $y$ | 7 | 22 | 1 | 16 | 19 | 10 | 13 | 4 |

**11** $y = 3(x - 2)^2 + 3$

**12 a** 8 m **b** 5 m **c** 3.34 m **d** 4.58 m

**13 a** I 300 cm$^2$  II 300 cm$^2$
III 600 cm$^2$  IV 55 cm$^2$

**b** 256 cm

## Exercise 6A  PAGE 117

**1** $x = -5, x = 3$
**2** $x = -8, x = -9$
**3** $x = 5.5, x = -5$
**4** $x = \pm 5$
**5** $x = \pm 7$
**6** $x = \pm 10$
**7** $x = -5, x = -4$
**8** $x = -5, x = 4$
**9** $x = 4, x = 5$
**10** $x = -4, x = 5$
**11** $x = -7, x = 5$
**12** $x = -3, x = -1$
**13** $x = -6, x = -1$
**14** $x = -7, x = -3$
**15** $x = -5, x = -3$
**16** $x = -2, x = 6$
**17** $x = -1, x = 5$
**18** $x = 0, x = 4$
**19** $x = -7, x = 2$
**20** $x = \pm 6$
**21** $x = -3$
**22** $x = -1, x = 4$
**23** $x = 4$
**24** $x = -5, x = 3$
**25** $x = 0, x = 3$
**26** $x = 3, x = 4$
**27** $x = -12, x = 2$
**28** $x = \pm 1.5$
**29** $x = \pm 0.2$
**30** $x = -3, x = 5$
**31** $x = 3$
**32** $x = 5$
**33** $x = 1.5, x = -4$
**34** $x = -4, x = \dfrac{2}{3}$

**35** $x = -1, x = 2.5$
**36** $x = -7, x = 0.2$
**37** $x = -3.5, x = 3$
**38** $x = \dfrac{2}{3}, x = 2.5$
**39** $x = 0.4, x = 0.5$
**40** The number is either –10 or 3.
**41** The number is –5.
**42** When the object hits the ground again $h = 0$ and $t = 8$.
**43** $t = 2$
**44** $p = -3$ or $p = 11$

## Exercise 6B  PAGE 123

**1** $x = -0.77, x = 0.43$
**2** $x = -2.30, x = 1.30$
**3** No real solutions
**4** $x = -2.82, x = -0.18$
**5** $x = -1.74, x = 0.34$
**6** $x = -1.47, x = 0.27$
**7** $t = 13.8$
**8** $p = 0.22$ or 2.78
**9** No real solutions
**10** Two real solutions
**11** No real solutions
**12** One real solution
**13** One real solution
**14** Two real solutions
**15** Two real solutions
**16** No real solutions
**17** One real solution
**18** $x \approx -2.7, x \approx 0.7$
**19** $x \approx -5.3, x \approx 1.3$
**20** $x \approx 0.4, x \approx 3.6$
**21** No real solutions
**22** $x \approx -5.7, x \approx -0.3$
**23** $x \approx -0.2, x \approx 4.2$
**24** $x = 2.13, x = 9.87$
**25** No real solutions
**26** $x = 7.87, x = 0.13$
**27** $x = -7.65, x = 0.65$
**28** $x = -4.19, x = 1.19$
**29** $x = 1, x = -1.5$
**30** $x = 1 \pm \sqrt{6}$
**31** $x = 3 \pm 2\sqrt{2}$
**32** $x = -5 \pm 4\sqrt{2}$
**33** $x = -\dfrac{5}{2} \pm \dfrac{\sqrt{35}}{2}$
**34** $x = -\dfrac{5}{6} \pm \dfrac{\sqrt{13}}{6}$
**35** $x = -\dfrac{1}{10} \pm \dfrac{\sqrt{21}}{10}$
**36** $x = 1.56, x = -2.56$
**37** $x = 4.11, x = -0.61$
**38** $x = 2.18, x = 0.15$
**39** $x = 4.41, x = 1.59$
**40** $x = 3.19, x = -2.19$
**41** $x = 0.76, x = -1.09$
**42** $x = -\dfrac{3}{2} \pm \dfrac{\sqrt{5}}{2}$
**43** $x = \dfrac{7}{2} \pm \dfrac{3\sqrt{5}}{2}$
**44** $x = -\dfrac{1}{4} \pm \dfrac{\sqrt{41}}{4}$
**45** $x = \dfrac{5}{6} \pm \dfrac{\sqrt{37}}{6}$
**46** $x = -\dfrac{1}{10} \pm \dfrac{\sqrt{101}}{10}$
**47** $x = -1 \pm \dfrac{\sqrt{2}}{2}$
**48** 2 real roots
**49** no real roots
**50** 2 real roots
**51** 2 real roots
**52** 1 real root
**53** no real roots

## Miscellaneous exercise six   PAGE 125

**1** The number could be –5 or it could be 3.

**2** A: $y = -x + 60$, B: $y = 60$, C: $y = 2x - 60$, D: $x = 60$,
E: $y = -2x + 30$, F: $y = 0.5x + 30$

**3 a** AD is of length 6 units. DB is of length 6 units.
The straight line through A and B has a gradient of 1.

   **b** DE is of length 8 units. EC is of length 4 units.
The straight line through D and C has a gradient of 0.5.

   **c** The straight line through D and F has a gradient of 0.75.

**4** $a = 1, b = 1, c = 2, d = 77, e = 77, f = -1$ or 3

**5** 15 cm by 2.4 cm.

**6** 11.2 cm$^2$

**7** A: $y = x^2 - 1$, B: $y = (x - 7)^2$, C: $y = (x + 9)^2 + 2$,
D: $y = (x + 5)^2 - 8$, E: $y = -(x - 4)^2 + 1$,
F: $y = 2(x - 10)^2$, G: $y = 4(x + 5)^2 - 3$, H: $y = -2(x + 10)^2$

**8** 8.3 cm

**9** 11.49

## Exercise 7A   PAGE 134

**1 a** $(0, 1)$    **b** $(0, -5)$    **c** $(0, 8)$

   **d** $(0, 6)$    **e** $(0, 2)$    **f** $(0, 3)$

**2 a** $(2, 0), (3, 0), (4, 0)$    **b** $(-7, 0), (1, 0), (5, 0)$

   **c** $(2.5, 0), (-1, 0), (0.6, 0)$   **d** $(1, 0), (-1, 0), (7, 0)$

   **e** $(0, 0), (0.25, 0), (3.5, 0)$   **f** $(-1, 0), (5, 0)$

   **g** $(-3, 0), (0, 0), (3, 0)$    **h** $(-5, 0), (0, 0), (3, 0)$

**3** $(2.20, 0)$

**4 a** $k = -6$    **b** $(-6, 0), (-2, 0), (3, 0)$

**5 a** $0$    **b** $0$

   **c** $-12$    **d** $0, (x - 6)(x + 1)(x - 1)$

**6 a** $-8$    **b** $0$

   **c** $0, (x - 2)(x - 3)(x - 5)$

**7 a** $a = 1, c = -5$    **b** $b = -4$

   **c** $(-1, 0), (\frac{2}{3}, 0), (5, 0)$

## Exercise 7B   PAGE 139

**1** B: $y = \sqrt{x - 3}$, C: $y = \sqrt{x} + 4$, D: $y = \sqrt{x + 3} - 5$

**2 a** $y = \frac{1}{x} + 1$    **b** $y = \frac{1}{x} + 2$    **c** $y = \frac{1}{x} - 1$

**3 a** $y = \frac{1}{x + 1}$    **b** $y = \frac{1}{x - 3}$    **c** $y = \frac{1}{x - 1}$

**4** The graph of $y = x^3 + 1$ is that of $y = x^3$ translated up 1 unit.

**5** The graph of $y = \frac{1}{x - 1}$ is that of $y = \frac{1}{x}$ translated 1 unit to the right.

**6** The graph of $y = 2\sqrt{x}$ is that of $y = \sqrt{x}$ dilated parallel to the $y$-axis, scale factor 2.

**7** The graph of $y = (x - 3)^2$ is that of $y = (x + 4)^2$ translated 7 units right.

**8** The graph of $y = \sqrt{x - 2} + 1$ is that of $y = \sqrt{x}$ translated 2 units right and 1 unit up.

**9** The graph of $y = \frac{3}{x - 1}$ is that of $y = \frac{1}{x}$ translated 1 unit to the right and dilated parallel to the $y$-axis, scale factor 3.

**10 a** B and F    **b** D

   **c** C, E, G and H    **d** H

   **e** $C \to E, G \to H$    **f** $A \to C, E \to G, H \to I$

**11** A(0, 10), B(–0.51, 0), C(3.08, 0), D(6.42, 0), E(1, 17),
F(5, –15), G(3, 1)

**12 a** When $P = 40$, $V = 10$.

   **b** When $P = 20$, $V = 20$.

   **c** Volume cannot be negative. With a non zero mass there must be some volume.
Thus $V > 0$ would be a suitable domain for $V$.

**13** $a = 4, b = 0.5, c = 4, d = 2, e = 3, f = 1, g = 3, h = -0.5, i = 3$
A(0, 8), B(–2, 0), C(0, 7), D(0, 2), E($\frac{2}{3}$, 0), F(–3, 0),
G(0, 9), H(0, 4), I(4, 0)

## Exercise 7C   PAGE 145

**1 a** Reflect in the $x$-axis.

   **b** Dilate parallel to the $x$-axis, scale factor 0.25.

   **c** Dilate parallel to the $y$-axis, scale factor 4.

**2 a** Reflect in the $x$-axis.

   **b** Translate 5 units down.

   **c** Dilate parallel to the $x$-axis, scale factor 2.

**3 a** Translate 3 units right.

   **b** Dilate parallel to $y$-axis scale factor 3.
(Or: Dilate parallel to $x$-axis scale factor $\frac{1}{\sqrt{3}}$)

   **c** Dilate parallel to $x$-axis scale factor $\frac{1}{3}$.
(Or: Dilate parallel to $y$-axis scale factor 9.)

**4 a**

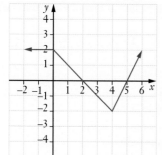

**b**

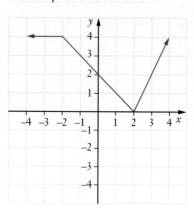

**c**

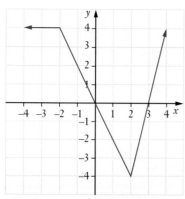

**d**

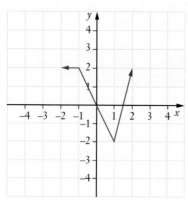

**e**

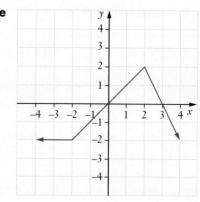

**f**

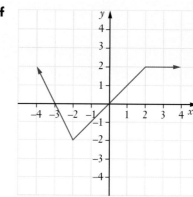

**5 a** 1    **b** 1.5    **c** 2    **d** 3

**e**

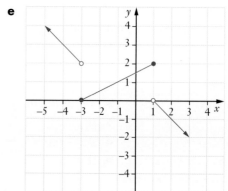

**f**

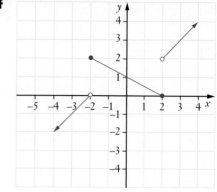

**g**

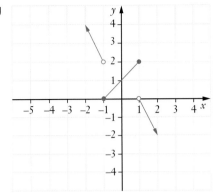

**h**

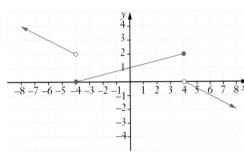

**i**

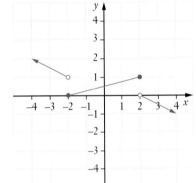

**6** A: III, B: X, C: IX, D: VI, E: I, F: II

**7 a** $(1, 0), (7, 0), (10, 0)$  **b** $(-1, 0), (2, 0), (3.5, 0)$
  **c** $(-2, 0), (4, 0), (7, 0)$  **d** $(-7, 0), (-4, 0), (2, 0)$
  **e** $(2, 8)$  **f** $(5, 1)$

## Exercise 7D   PAGE 149

**1** A, C, D

**2** $x^2 + y^2 = 100, a = 8, b = \sqrt{91}, c = -10, d = -5\sqrt{3}$

**3 a** $(x - 2)^2 + (y + 3)^2 = 25$
  **b** $(x - 3)^2 + (y - 2)^2 = 49$
  **c** $(x + 10)^2 + (y - 2)^2 = 45$
  **d** $(x + 1)^2 + (y + 1)^2 = 36$

**4 a** $x^2 + y^2 - 6x - 10y = -9$
  **b** $x^2 + y^2 + 4x - 2y = 2$
  **c** $x^2 + y^2 + 6x + 2y = -6$
  **d** $x^2 + y^2 - 6x - 16y = -45$

**5 a** $5, (0, 0)$  **b** $0.6, (0, 0)$  **c** $5, (3, -4)$
  **d** $10, (-7, 1)$  **e** $3, (3, -2)$  **f** $5, (-1, 3)$
  **g** $10, (-1, 7)$  **h** $15, (-5, 7)$  **i** $12, (10, 5)$
  **j** $2, (0.5, -2.5)$

**6** $\sqrt{5}$         **7** $y = -11x + 29$

**8** $(x - 3)^2 + (y - 4)^2 = 36$    **9** $(x + 4)^2 + (y + 3)^2 = 9$

**10 a** $(y - 2)^2 = x$      **b** $y^2 = x + 4$
  **c** $(y - 1)^2 = x - 2$    **d** $(y + 2)^2 = x - 3$

**11 a** 15
  **b** The circles have just one point in common
    because the distance between the centres equals
    the sum of the radii.

**12 a** $2\sqrt{5}$
  **b** The circles have no points in common because the
    distance between the centres exceeds the sum of
    the radii.

**13** $(1, -2)$ and $(8, 5)$      **14** $(-2, 7)$ and $(-10, 5)$

**15** $(5, 10)$           **16** $a < 26$

## Miscellaneous exercise seven   PAGE 152

**1 a** $y = 0.5(x + 3)(x - 2)(x - 4)$
  **b** $y = 2(x + 2)^2(x - 4)$

**2 a** $x = 2 \pm \sqrt{10}$       **b** $x = 2 \pm \sqrt{10}$

**3** Centre at $(-3, 5)$, radius 7

**4 a** 4    **b** 16    **c** 64    **d** 0 and 1

**5** $f_2$ gradient 2.5, $f_4$ gradient $-2$

**6** $y = 0.4x - 7$

**7 a** $x = -7, x = 2.25, x = 2.5$
  **b** $x = -5.25, x = -1.5, x = 7$
  **c** $x = 3$
  **d** No real solutions.

**8 a** Statements A and C   **b** Statements B and D
  **c** Statements B and D   **d** Statements A and C
  **e** Statement A        **f** Statements A and C
  **g** Statements B and D   **h** Statement B

**9 a** $x = -9, x = 3.5$    **b** $x = 2, x = 6$
  **c** $x = -1, x = 0.6$    **d** $x = -11, x = 0.8, x = 7$
  **e** $x = -5, x = 1, x = 3$   **f** $x = -5, x = -2, x = 1.5$

**10 a** Cubic      **b** Quadratic
   **c** None of the listed types    **d** Cubic
   **e** Reciprocal      **f** Linear

**11 a** $c = 4$    **b** $b = -5$    **c** $(x-3)(x-1)(x-4)$

**12 a** If $x$ is doubled in value $m$ must be halved in value if the system is to remain in balance.

   **b** The relationship between $x$ and $m$ is one of inverse proportion.

      If $x$ is multiplied by some factor $k$ then $m$ needs to be multiplied by $\dfrac{1}{k}$.

   **c** If $m = 20$ then $x$ must be 0.5 for the system to balance.

   **d** For the system to balance $x$ cannot take negative values, it cannot be zero and, from the length of the beam, $x$ cannot exceed 3. Hence the domain consists of all real numbers greater than 0 and less than or equal to 3, i.e. $\{x \in \mathbb{R}: 0 < x \le 3\}$.

      For this domain the rule will output $m$ values such that $m \ge \dfrac{10}{3}$. Hence the range consists of all real numbers greater than or equal to $\dfrac{10}{3}$, i.e. $\{m \in \mathbb{R}: m \ge \dfrac{10}{3}\}$.

**13** The triangular piece that has been removed has an area of 752 mm$^2$ and a perimeter of 128 mm, both answers given to the nearest whole number.

## Exercise 8A    PAGE 162

| | | | | | | | |
|---|---|---|---|---|---|---|---|
| **1** | 4 | **2** | 3 | **3** | 6 | **4** | 5 |
| **5** | 4 | **6** | 4 | **7** | 3, 6 | **8** | 3, 2 |
| **9** | 4, 5 | **10** | 2, 3 | **11** | 3, 2.5 | **12** | 1, 4 |

## Investigation    PAGE 163

In the graph of $y = a \sin x$ the amplitude is $a$, or to be more correct $|a|$.

Changing the value of $a$ changes the amplitude. The graph is stretched (or compressed) vertically. (If $a$ changes sign the graph reflects in the $x$-axis.)

The graph of $y = a \sin bx$ performs $b$ cycles in the interval that $y = \sin x$ would perform 1 cycle.

The period of the graph is $\dfrac{2\pi}{b}$, if radians are used, or $\dfrac{360}{b}$ for degrees.

Changing the value of $b$ changes the period. The graph is stretched or compressed horizontally.

In the graph of $y = a \sin [b(x - c)]$ changing the value of $c$ translates the graph horizontally.

In the graph of $y = a \sin [b(x - c)] + d$ changing the value of $d$ translates the graph vertically.

Discuss your findings for the cosine and tangent function with others in your class.

## Exercise 8B    PAGE 165

| | | | | | | | |
|---|---|---|---|---|---|---|---|
| **1 a** | 1 | **b** | 2 | **c** | 4 | **d** | 3 |
| **e** | 2 | **f** | 3 | **g** | 5 | **h** | 3 |
| **2 a** | 360° | **b** | 180° | **c** | 360° | **d** | 180° |
| **e** | 720° | **f** | 120° | **g** | 90° | **h** | 1080° |
| **i** | 180° | | | | | | |

| | | | | | | | |
|---|---|---|---|---|---|---|---|
| **3 a** | $2\pi$ | **b** | $\pi$ | **c** | $2\pi$ | **d** | $\dfrac{\pi}{2}$ |
| **e** | $\dfrac{\pi}{3}$ | **f** | $\dfrac{2\pi}{3}$ | **g** | $4\pi$ | **h** | $\pi$ |
| **i** | 0.5 | | | | | | |

**4 a** Max at $(\dfrac{\pi}{2}, 1)$. Min at $(\dfrac{3\pi}{2}, -1)$.

   **b** Max at $(\dfrac{\pi}{2}, 3)$. Min at $(\dfrac{3\pi}{2}, 1)$.

   **c** Max at $(\dfrac{3\pi}{2}, 1)$. Min at $(\dfrac{\pi}{2}, -1)$.

   **d** Max at $(\dfrac{\pi}{4}, 4)$ and at $(\dfrac{5\pi}{4}, 4)$.

      Min at $(\dfrac{3\pi}{4}, 2)$ and at $(\dfrac{7\pi}{4}, 2)$.

   **e** Max at $(\dfrac{3\pi}{4}, 4)$. Min at $(\dfrac{7\pi}{4}, 2)$.

| | | | | | | | |
|---|---|---|---|---|---|---|---|
| **5 a** | 3, 90° | **b** | 2, 120° | **c** | 2, 60° | **d** | 3, 270° |
| **6 a** | $3, \dfrac{\pi}{4}$ | **b** | $5, \dfrac{3\pi}{2}$ | **c** | $2, \dfrac{11\pi}{6}$ | **d** | $3, \dfrac{\pi}{6}$ |
| **7 a** | 2 | | | **b** | 3 | | |
| **c** | $-3$ | | | **d** | Approx. $-1.3$ | | |
| **8 a** | 3 | | | **b** | $-2$ | | |
| **9 a** | 2 | | | **b** | $-1$ | | |
| **10 a** | 2, 3 | **b** | $-3, 2$ | **c** | 2, 6 | **d** | $3, \dfrac{2\pi}{3}$ |
| **11 a** | 1, 2 | **b** | $-3, 3$ | **c** | $-3, 2$ | **d** | $2, \dfrac{\pi}{2}$ |

**12 a** $a = 2, b = 30, 390$    **b** $y = -2 \sin (x - 210)°$

**13 a** Period 2, Amplitude 3.

**b**

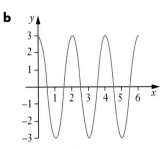

**14 a** Period 4, Amplitude 5.

**b**

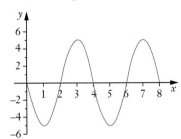

**15**

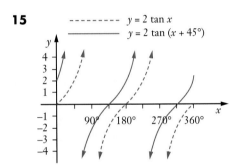

$y = 2 \tan x$ (dashed)
$y = 2 \tan (x + 45°)$ (solid)

**16**

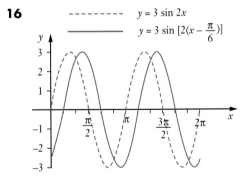

$y = 3 \sin 2x$ (dashed)
$y = 3 \sin [2(x - \frac{\pi}{6})]$ (solid)

## Exercise 8C   PAGE 172

| | | |
|---|---|---|
| **1** Positive | **2** Positive | **3** Negative |
| **4** Negative | **5** Negative | **6** Negative |
| **7** Positive | **8** Positive | **9** Positive |
| **10** Negative | **11** Negative | **12** Negative |
| **13** $\sin 40°$ | **14** $-\sin 70°$ | **15** $-\sin 20°$ |
| **16** $\sin 80°$ | **17** $\sin \frac{\pi}{6}$ | **18** $-\sin \frac{\pi}{6}$ |

**19** $\sin \frac{\pi}{5}$   **20** $-\sin \frac{\pi}{5}$   **21** $-\cos 80°$

**22** $-\cos 20°$   **23** $\cos 60°$   **24** $\cos 60°$

**25** $-\cos \frac{\pi}{5}$   **26** $-\cos \frac{\pi}{10}$   **27** $-\cos \frac{\pi}{10}$

**28** $\cos \frac{\pi}{10}$   **29** $-\tan 80°$   **30** $\tan 20°$

**31** $-\tan 60°$   **32** $\tan 20°$   **33** $\tan \frac{\pi}{5}$

**34** $-\tan \frac{\pi}{5}$   **35** $\tan \frac{\pi}{5}$   **36** $-\tan \frac{\pi}{5}$

**37** $-\frac{\sqrt{3}}{2}$   **38** $\frac{1}{\sqrt{3}}$   **39** $-\frac{1}{2}$

**40** $0$   **41** $0$   **42** $\frac{\sqrt{3}}{2}$

**43** $-\frac{1}{\sqrt{2}}$   **44** $-\frac{1}{\sqrt{2}}$   **45** $-\frac{1}{2}$

**46** $-\frac{\sqrt{3}}{2}$   **47** $\frac{1}{\sqrt{3}}$   **48** $-\frac{1}{\sqrt{2}}$

**49** $\frac{1}{\sqrt{2}}$   **50** $0$   **51** $1$

**52** $\frac{1}{2}$

## Exercise 8D   PAGE 177

**1** $60°, 300°$   **2** $210°, 330°$   **3** $45°, 225°$

**4** $225°, 315°$   **5** $\frac{\pi}{4}, \frac{3\pi}{4}$   **6** $\frac{3\pi}{4}, \frac{5\pi}{4}$

**7** $\frac{3\pi}{4}, \frac{7\pi}{4}$   **8** $\frac{\pi}{3}, \frac{4\pi}{3}$   **9** $\pm 30°$

**10** $-90°$   **11** $-30°, 150°$   **12** $0°, \pm 180°$

**13** $\frac{\pi}{3}, \frac{2\pi}{3}$   **14** $\pm \frac{2\pi}{3}$   **15** $\frac{\pi}{6}, \frac{5\pi}{6}$

**16** $\pm \frac{\pi}{2}$   **17** $\pi + 0.98$   **18** $\pm 116.1°$

**19** $15°, 105°$   **20** $\frac{\pi}{24}, \frac{11\pi}{24}, \frac{13\pi}{24}, \frac{23\pi}{24}$

**21** $-70°, 10°, 50°$   **22** $\frac{\pi}{6}, \frac{\pi}{3}, \frac{7\pi}{6}, \frac{4\pi}{3}$

**23** $\frac{5\pi}{18}, \frac{7\pi}{18}, \frac{17\pi}{18}, \frac{19\pi}{18}, \frac{29\pi}{18}, \frac{31\pi}{18}$

**24** $\frac{\pi}{6}, \frac{5\pi}{6}, \frac{3\pi}{2}$   **25** $45°, 135°, 225°, 315°$

**26** $\pm \frac{\pi}{6}, \pm \frac{5\pi}{6}$   **27** $0°, \pm 60°, \pm 180°$

**28** $\pm\dfrac{\pi}{3}, \pm\pi$      **29** $\dfrac{5\pi}{12}, \dfrac{23\pi}{12}$

## Exercise 8E   PAGE 180

**1** $14.5°, 165.5°$      **2** $\pm\dfrac{5\pi}{6}, \pm\dfrac{\pi}{6}$

**3** $\dfrac{\pi}{2}$      **4** $\dfrac{\pi}{6}, \dfrac{\pi}{2}, \dfrac{5\pi}{6}, \dfrac{3\pi}{2}$

**5** $0°, 180°, 210°, 330°, 360°$

**6** $11.5°, 120°, 168.5°, 240°$

**7** $\dfrac{\pi}{3}, \dfrac{2\pi}{3}, \dfrac{4\pi}{3}, \dfrac{5\pi}{3}$

**8** $-135°, -63.4°, 45°, 116.6°$

**9** $\pm60°$

**10** $\dfrac{\pi}{6}, \dfrac{\pi}{2}, \dfrac{5\pi}{6}, \dfrac{13\pi}{6}, \dfrac{5\pi}{2}, \dfrac{17\pi}{6}$

## Exercise 8F   PAGE 185

**1** $\sin 3x$    **2** $\cos 2x$    **3** $\sin 4x$

**4** $\cos 8x$    **5** $\dfrac{\sqrt{2}(\sqrt{3}+1)}{4}$    **6** $2 - \sqrt{3}$

**7** $\dfrac{\sqrt{2}(\sqrt{3}+1)}{4}$    **8** $\dfrac{\sqrt{2}(\sqrt{3}-1)}{4}$    **9** $2 + \sqrt{3}$

**10** $a = \sqrt{2}, b = \sqrt{2}$      **11** $c = 4\sqrt{3}, d = 4$

**12** $e = 2\sqrt{3}, f = -2$      **13** $\sqrt{3}, \dfrac{4\pi}{3}$

**14 a** $\dfrac{56}{65}$      **b** $\dfrac{63}{65}$

**15 a** $-\dfrac{44}{125}$      **b** $\dfrac{3}{5}$

**21 a** $-\dfrac{56}{65}$      **b** $\dfrac{63}{65}$      **c** $-\dfrac{56}{33}$

**22** $\dfrac{\pi}{12}, \dfrac{7\pi}{12}$      **23** $80°, 320°$

**24** $-40°, 80°$      **25** $30°, 210°$

## Alternating currents   PAGE 186

Amplitude $\approx 340$ volts. Period 0.02 seconds. $V = 340 \sin 100\pi t$

## Average weekly temperatures   PAGE 187

$T = 18 - 12 \sin \dfrac{\pi}{26} x$

Average weekly temperature exceeded 25°C on 15 of the weeks.

## Tidal motion   PAGE 187

Compare your answers to those of others in your class.

## Miscellaneous exercise eight   PAGE 188

**1** Amplitude 5, Period $2\pi$.

**2** Amplitude 7, Period $2\pi$.

**3** Amplitude 3, Period $2\pi$.

**4** Amplitude 1, Period $\dfrac{2\pi}{2}$, i.e. $\pi$.

**5** Amplitude 1, Period $\dfrac{2\pi}{3}$.

**6** Amplitude 1, Period $\dfrac{2\pi}{0.5}$, i.e. $4\pi$.

**7** Amplitude 3, Period $\dfrac{2\pi}{4}$, i.e. $\dfrac{\pi}{2}$.

**8** Amplitude 4, Period $\dfrac{2\pi}{5}$.

**9** Amplitude 2, Period $\dfrac{2\pi}{\pi}$. i.e. 2.

**10**

| $\theta$ | $-\dfrac{3\pi}{4}$ | $-\dfrac{2\pi}{3}$ | $\dfrac{\pi}{6}$ | $\dfrac{\pi}{4}$ | $\dfrac{4\pi}{3}$ | $\dfrac{7\pi}{3}$ | $\dfrac{9\pi}{4}$ | $11\pi$ |
|---|---|---|---|---|---|---|---|---|
| $\sin\theta$ | $-\dfrac{\sqrt{2}}{2}$ | $-\dfrac{\sqrt{3}}{2}$ | $\dfrac{1}{2}$ | $\dfrac{\sqrt{2}}{2}$ | $-\dfrac{\sqrt{3}}{2}$ | $\dfrac{\sqrt{3}}{2}$ | $\dfrac{\sqrt{2}}{2}$ | $0$ |
| $\cos\theta$ | $-\dfrac{\sqrt{2}}{2}$ | $-\dfrac{1}{2}$ | $\dfrac{\sqrt{3}}{2}$ | $\dfrac{\sqrt{2}}{2}$ | $-\dfrac{1}{2}$ | $\dfrac{1}{2}$ | $\dfrac{\sqrt{2}}{2}$ | $-1$ |
| $\tan\theta$ | $1$ | $\sqrt{3}$ | $\dfrac{\sqrt{3}}{3}$ | $1$ | $\sqrt{3}$ | $\sqrt{3}$ | $1$ | $0$ |

**11 a** Neither    **b** Parallel    **c** Perpendicular

**12 a** 13.2      **b** 13.2

**13** The smallest angle of the triangle is 21°, to the nearest degree.

**14 a** $(2, 0), (3, 0), (-2, 0), (-7, 0)$

    **b** $(0, 0), (2, 0), (-3, 0), (4, 0)$

    **c** $(2, 0), (3, 0), (-3, 0)$

    **d** $(2, 0)$

    **e** $(7, 0)$

    **f** $(-5, 0), (-1.5, 0), (3.5, 0), (6, 0)$

**15** $k_1 = 0.5, k_2 = -10, k_3 = 4, k_4 = 2, k_5 = 2, k_6 = -12, k_7 = 13,$
$k_8 = 10, k_9 = 1, k_{10} = -3, k_{11} = 1, k_{12} = 2, k_{13} = -3,$
$k_{14} = -5, k_{15} = 1, k_{16} = -9, k_{17} = 27, k_{18} = -26, k_{19} = 5,$
$k_{20} = 2, k_{21} = 5, k_{22} = 2, k_{23} = 45, k_{24} = 5, k_{25} = -4, k_{26} = 3$

**16 a** $a = 2, c = -12$      **b** $b = 5$

    **c** $(-4, 0), (1.5, 0), (2, 0)$

**17**
**a** Maximum turning point at $(-1, 26)$. Minimum turning point at $(3, -6)$.

**b** Maximum turning point at $(-1, 16)$. Minimum turning point at $(3, -16)$.

**c** Maximum turning point at $(1, 21)$. Minimum turning point at $(-3, -11)$.

**d** Maximum turning point at $(3, 11)$. Minimum turning point at $(-1, -21)$.

**e** Maximum turning point at $(-1, 63)$. Minimum turning point at $(3, -33)$.

**f** Maximum turning point at $(-0.5, 21)$. Minimum turning point at $(1.5, -11)$.

**18** **a** $(-1, 0), (2, 0), (5, 0)$ **b** $(0, 10)$

**c** $a = 8$ **d** $b = -8$

**e** $c = -10$ **f** $(3, -4)$. Minimum

**g** $(0, 5)$ **h** $d = 0$

**i** Use a graphic calculator to check your sketch. $x \approx -0.4, x \approx 2.4, x = 5$

**19** $269.4 \text{ cm}^2$

## Exercise 9A   PAGE 193

**1 a** $\dfrac{1}{2}$ **b** $\dfrac{1}{2}$ **c** $\dfrac{1}{2}$

**d** $\dfrac{1}{3}$ **e** $\dfrac{2}{3}$

**2 a** $\dfrac{1}{2}$ **b** $\dfrac{1}{2}$ **c** $\dfrac{5}{12}$

**d** $\dfrac{1}{18}$ **e** $\dfrac{5}{18}$ **f** $\dfrac{13}{18}$

**3 a** $0.3$ **b** $0.7$ **c** $0.5$ **d** $0.8$

**4 a** $\dfrac{1}{8}$ **b** $\dfrac{1}{8}$ **c** $\dfrac{3}{8}$ **d** $\dfrac{1}{2}$

**e** $\dfrac{1}{8}$ **f** $\dfrac{1}{4}$

**5 a** $0$ **b** $\dfrac{5}{6}$ **c** $\dfrac{1}{3}$

**d** $\dfrac{1}{3}$ **e** $\dfrac{1}{6}$

**6 a** $\dfrac{1}{10}$ **b** $\dfrac{3}{20}$ **c** $\dfrac{47}{100}$ **d** $\dfrac{12}{25}$

**e** $\dfrac{7}{10}$ **f** $\dfrac{11}{50}$

**7 a** $0.327$ **b** $0.672$

**8 a** $7$ **b** $9$ **c** $10$ **d** $3$

**e** $\{8, 9, 10\}$ **f** $\{1, 3, 5, 7, 9\}$

**g** $\{9\}$ **h** $\{1, 3, 5, 7, 8, 9, 10\}$

**9 a** $67$ **b** $3$

**10** $17$ **11** $17$ **12** $11$

**13 a** $65$ **b** $12$ **c** $8$

**14 a** $0.6$ **b** $0.2$ **c** $0.1$ **d** $0.7$

**e** $0.9$ **f** $0.3$

**15 a** $\dfrac{10}{19}$ **b** $\dfrac{12}{19}$ **c** $\dfrac{9}{19}$ **d** $\dfrac{10}{19}$

**e** $\dfrac{1}{19}$ **f** $\dfrac{3}{19}$ **g** $\dfrac{13}{19}$ **h** $\dfrac{16}{19}$

**i** $\dfrac{18}{19}$

**16 a** $\dfrac{17}{40}$ **b** $\dfrac{13}{40}$ **c** $0$ **d** $\dfrac{3}{4}$

**e** $\dfrac{1}{4}$ **f** $1$

**17** $0.2$

**18 a** $0.3$ **b** $0.38$

**19 a** $\dfrac{13}{20}$ **b** $\dfrac{13}{15}$

## Exercise 9B   PAGE 199

**1 a** $\dfrac{1}{6}$ **b** $\dfrac{1}{5}$

**2 a** $\dfrac{1}{4}$ **b** $\dfrac{1}{3}$

**3 a** $\dfrac{1}{18}$ **b** $\dfrac{1}{6}$

**4 a** $\dfrac{1}{52}$ **b** $\dfrac{1}{20}$

**5 a** $\dfrac{7}{25}$ **b** $\dfrac{15}{26}$

**6 a** $\dfrac{3}{5}$ **b** $\dfrac{47}{100}$ **c** $\dfrac{7}{10}$ **d** $\dfrac{37}{100}$

**e** $\dfrac{2}{5}$ **f** $\dfrac{53}{100}$ **g** $\dfrac{37}{47}$ **h** $\dfrac{23}{53}$

**7 a** $\dfrac{1}{2}$ **b** $\dfrac{1}{3}$ **c** $\dfrac{1}{3}$ **d** $1$

**e** $1$ **f** $\dfrac{1}{2}$

**8 a** $0.7$ **b** $0.3$ **c** $0.9$ **d** $0.3$

**e** $0.7$ **f** $\dfrac{1}{3}$ **g** $\dfrac{1}{7}$ **h** $\dfrac{7}{9}$

**i** $1$

**9 a** $\dfrac{1}{3}$ **b** $\dfrac{5}{9}$ **c** $\dfrac{7}{9}$ **d** $\dfrac{1}{9}$

**e** $\frac{2}{3}$    **f** $\frac{4}{9}$    **g** $\frac{1}{5}$    **h** $\frac{1}{3}$

**i** $\frac{5}{7}$

**10 a** $\frac{2}{5}$    **b** $\frac{1}{10}$    **c** $\frac{1}{4}$    **d** $\frac{3}{5}$

**e** $\frac{1}{3}$

**11 a** $\frac{1}{6}$    **b** $\frac{1}{3}$    **c** $0$    **d** $0$

**e** $\frac{1}{3}$    **f** $\frac{2}{3}$    **g** $\frac{1}{3}$    **h** $\frac{2}{3}$

**12 a** $\frac{1}{10}$    **b** $\frac{1}{5}$    **c** $0$    **d** $\frac{1}{4}$

**e** $\frac{1}{6}$    **f** $\frac{5}{6}$    **g** $\frac{2}{5}$

**13 a** $\frac{1}{9}$    **b** $0$    **c** $\frac{1}{6}$    **d** $\frac{1}{6}$

**e** $\frac{4}{9}$

**14 a** $\frac{1}{6}$    **b** $\frac{1}{3}$    **c** $\frac{1}{6}$    **d** $\frac{1}{5}$

**e** $\frac{1}{9}$    **f** $\frac{1}{2}$

**15 a** $\frac{1}{20}$    **b** $\frac{1}{2}$    **c** $\frac{2}{5}$    **d** $\frac{3}{10}$

**e** $\frac{3}{10}$    **f** $\frac{1}{10}$    **g** $\frac{1}{5}$    **h** $\frac{1}{11}$

**i** $\frac{1}{6}$    **j** $\frac{1}{2}$    **k** $\frac{1}{5}$    **l** $\frac{1}{3}$

**16 a** $\frac{1}{2}$    **b** $\frac{1}{4}$    **c** $1$    **d** $\frac{1}{2}$

**e** $\frac{1}{3}$    **f** $0$

**17 a** $\frac{1}{2}$    **b** $\frac{1}{2}$

## Exercise 9C    PAGE 204
(Tree diagrams not shown here.)

**1 a** $\frac{1}{2}$    **b** $\frac{2}{3}$    **c** $\frac{3}{4}$    **d** $\frac{1}{2}$

**2 a** $\frac{1}{3}$    **b** $\frac{5}{9}$    **c** $\frac{1}{3}$    **d** $\frac{2}{3}$

**3 a** $\frac{1}{9}$    **b** $\frac{5}{9}$    **c** $\frac{2}{3}$    **d** $\frac{1}{2}$

**4 a** $\frac{1}{4}$    **b** $\frac{3}{16}$    **c** $\frac{2}{3}$    **d** $\frac{6}{13}$

**5 a** $\frac{1}{20}$    **b** $\frac{1}{5}$    **c** $\frac{1}{5}$    **d** $\frac{1}{20}$

**e** $\frac{1}{4}$    **f** $\frac{1}{4}$    **g** $\frac{1}{4}$

## Exercise 9D    PAGE 206

**1 a** $\frac{5}{6}$    **b** $\frac{1}{36}$    **c** $\frac{11}{36}$    **d** $\frac{1}{18}$

**e** $\frac{4}{9}$

**2 a** $\frac{1}{2}$    **b** $\frac{1}{6}$    **c** $\frac{1}{12}$    **d** $\frac{7}{12}$

**e** $\frac{1}{4}$    **f** $\frac{1}{2}$

**3 a** $\frac{1}{52}$    **b** $\frac{1}{13}$    **c** $\frac{1}{2}$    **d** $\frac{1}{13}$

**e** $\frac{1}{4}$    **f** $\frac{1}{13}$    **g** $\frac{12}{13}$    **h** $\frac{3}{13}$

**i** $\frac{1}{26}$    **j** $\frac{7}{13}$    **k** $\frac{1}{52}$    **l** $\frac{4}{13}$

**4 a** $\frac{1}{2}$    **b** $\frac{1}{4}$    **c** $\frac{3}{4}$    **d** $\frac{1}{2}$

## Exercise 9E    PAGE 214

**1 a** $0.6$    **b** $0.12$    **c** $0.6$    **d** $0.3$
**e** $0.2$    **f** $0.7$

**2 a** $0.4$    **b** $0.2$    **c** $0.26$    **d** $0.8$

**e** $0.1$    **f** $\frac{3}{13}$

**3 a** $\frac{7}{30}$    **b** $\frac{7}{15}$    **c** $\frac{8}{15}$    **d** $\frac{7}{8}$

**4 a** $\frac{3}{4}$    **b** $\frac{4}{5}$    **c** $\frac{1}{5}$

**5 a** $0.6$    **b** $0.56$    **c** $\frac{4}{7}$    **d** $\frac{9}{11}$

**6 a** $\frac{1}{2}$    **b** $\frac{1}{6}$    **c** $\frac{17}{60}$    **d** $\frac{43}{60}$

**e** $\frac{1}{6}$    **f** $\frac{4}{5}$    **g** $\frac{10}{17}$    **h** $\frac{15}{43}$

**7 a** $\frac{25}{36}$    **b** $\frac{16}{25}$

**8 a** $\frac{4}{9}$    **b** $\frac{3}{4}$    **c** $\frac{55}{83}$

**9 a** 0.912 **b** 0.038 **c** 0.039 **d** 0.563

**10 a** $\dfrac{3}{5}$ **b** $\dfrac{1}{6}$ **c** $\dfrac{4}{15}$ **d** $\dfrac{1}{6}$

**e** $\dfrac{5}{18}$

## Exercise 9F PAGE 223

**1 a** $\dfrac{1}{3}$ **b** $\dfrac{7}{15}$ **c** $\dfrac{1}{3}$ **d** $\dfrac{2}{3}$

**2 a** 0.8472 **b** 0.1528

**3 a** 0.9265 **b** 0.0735

**4 a** $\dfrac{1}{30}$ **b** $\dfrac{1}{20}$ **c** $\dfrac{1}{6}$ **d** $\dfrac{5}{6}$

**5 a** 0.064 **b** 0.118 **c** 0.216 **d** 0.784

**6 a** 0.1248 **b** 0.1152

**7** $\dfrac{5}{12}$

**8** 0.01097

**9 a** $\dfrac{1}{2}$ **b** $\dfrac{2}{3}$ **c** $\dfrac{1}{3}$ **d** $\dfrac{5}{6}$

**10 a** $\dfrac{1}{3}$ **b** $\dfrac{1}{2}$ **c** $\dfrac{1}{6}$ **d** $\dfrac{2}{3}$

**11 a** 0.8 **b** 0.2 **c** 0.2 **d** 0.25

**12 a** 0.4 **b** 0.5 **c** 0.8

**13 a** 0.00000003 **b** 0.66 **c** 0.34

**14 a** $\dfrac{1}{36}$ **b** $\dfrac{25}{36}$ **c** $\dfrac{11}{36}$

**15 a** 0.00005 **b** 0.98505 **c** 0.01495

**16 a** 0.0000001 **b** 0.983 **c** 0.017

**17 a** $\dfrac{1}{2}$ **b** $\dfrac{2}{5}$ **c** $\dfrac{3}{5}$ **d** $\dfrac{7}{10}$

**18 a** $\dfrac{8}{15}$ **b** $\dfrac{4}{5}$

**19 a** 0.13 **b** 0.2

**20** 0.8 **21** $\dfrac{3}{4}$

## Exercise 9G PAGE 228

**1** Dependent **2** Independent
**3** Independent **4** Dependent
**5** Mutually exclusive **6** Mutually exclusive
**7** Not mutually exclusive **8** Not mutually exclusive
**9** **a** and **d**
**10 a** Independent **b** Dependent
**12 a** 0.2 **b** 0.25 **c** 0.05 **d** 0.6

**13 a** 0 **b** 0 **c** 0 **d** 0.5
**15 a** 0.3 **b** 0.8 **c** 0.6 **d** 0.5
**16** 0.8
**17** 0.2
**18 a** 0.4 **b** 0.48
**19 a** 0.3 **b** 0.4
**20 a** 0 **b** 15
**21 a** 0.35 (2 dp) **b** 0.45 (2 dp) **c** 0.19 (2 dp)

The disparity between the three probabilities suggests that being an Engineering student is **not** independent of gender. Whilst 35% of all students at the college are Engineering students, for males at the college this rises to 45% whilst for females it is just 19%.

**22 a** 0.35 (2 dp) **b** 0.34 (2 dp) **c** 0.35 (2 dp)

The fact that the three probabilities are almost identical suggests that whether the student is on the honours course is independent of gender. The proportion of students on the honours course is almost exactly the same whether we are considering all the students, just the males or just the females.

## Miscellaneous exercise nine PAGE 231

**1 a** 0.5 **b** 0.5 **c** $\dfrac{5}{6}$ **d** $\dfrac{1}{6}$

**e** 0.5

**3** $x = -4, x = 4.5$

**4 a** 1 **b** $\dfrac{1}{3}$ **c** $\dfrac{1}{2}$

**5 a** Period 180°, amplitude 4 units.
**b** Period 120°, amplitude 3 units.

**6** $x = -148, x = -32, x = 212, x = 328.$

**7** C has coordinates (7, 1)

**8 a** $\dfrac{55}{93}$ **b** $\dfrac{7}{93}$ **c** $\dfrac{7}{22}$ **d** $\dfrac{7}{38}$

**9 a** (0, –28) **b** A(–2, 0), D(7, 0)
**c** $a = 4, b = -108$ **d** $-108 < p < 0$

**10** $\dfrac{\pi}{3}, \dfrac{14\pi}{13}, \dfrac{5\pi}{3}, \dfrac{25\pi}{13}$

**11 a** $\dfrac{1}{15}$ **b** $\dfrac{3}{10}$ **c** $\dfrac{5}{14}$ **d** $\dfrac{5}{7}$

**12 a** 0.32 **b** 0.2 **c** 0.68 **d** 0.2
**e** 0.6
**f** Yes. Justification: $P(B|A) = 0.2 = P(B)$
[or $P(A|B) = 0.6 = P(A)$]
[or $P(A \cap B) = 0.12 = P(A)\,P(B)$]

## Exercise 10A  PAGE 238

**1** 40 320　**2** 48　**3** 10　**4** 90

**5** 90　**6** 56　**7** 8　**8** 970 200

**9** 96　**10** 120, 5!

**11** 20, $\dfrac{5!}{5-2!}$ i.e. $\dfrac{5!}{3!}$　**12** 60, $\dfrac{5!}{2!}$

**13** 650, $\dfrac{26!}{24!}$　**14** 358 800, $\dfrac{26!}{22!}$

**15** 40 320, 8!　**16** 504, $\dfrac{9!}{6!}$

## Exercise 10B  PAGE 242

**1** 330　**2** 18 564

**3** 210　**4** 455

**5** 792　**6** 133 784 560

**7** 5 245 786　**8** 3003, 10

## Exercise 10C  PAGE 243

**1** $a^8 + 8a^7b + 28a^6b^2 + 56a^5b^3 + 70a^4b^4 + 56a^3b^5 + 28a^2b^6 + 8ab^7 + b^8$

**2** $a^{10} + 10a^9b + 45a^8b^2 + 120a^7b^3 + 210a^6b^4 + 252a^5b^5 + 210a^4b^6 + 120a^3b^7 + 45a^2b^8 + 10ab^9 + b^{10}$

**3** $x^8 - 8x^7y + 28x^6y^2 - 56x^5y^3 + 70x^4y^4 - 56x^3y^5 + 28x^2y^6 - 8xy^7 + y^8$

**4** $x^6 + 12x^5y + 60x^4y^2 + 160x^3y^3 + 240x^2y^4 + 192xy^5 + 64y^6$

**5** $p^6 - 12p^5q + 60p^4q^2 - 160p^3q^3 + 240p^2q^4 - 192pq^5 + 64q^6$

**6** $243x^5 - 810x^4y + 1080x^3y^2 - 720x^2y^3 + 240xy^4 - 32y^5$

## Miscellaneous exercise ten  PAGE 246

**1 a** $x = 0.625$　**b** $x = -4$

　**c** $x = -1$　**d** $x = 1.4$

　**e** $x = 3, x = -2$　**f** $x = 1, x = -5$

　**g** $x = 0.5, x = -7$　**h** $x = -3, x = 0.25, x = 1.8$

　**i** $x = 9, x = -3$　**j** $x = -2, x = 3.5$

　**k** $x = -1, x = 0, x = 6$　**l** $x = -0.8, x = 1.5$

**2** A: $x = -6$, B: $y = 16$, C: $y = 4x$, D: $y = 2x$, E: $y = x$, F: $y = x - 5$, G: $y = -2x$, H: $y = -4x + 20$

**3** 0.1

**4** The number could be −3.5 or it could be 1.5.

**5** The fire is approximately 19.7 km from lookout No.1 and 18.8 km from lookout No.2.

**6** $\dfrac{1}{6}$

**7 a** $y = 3x + 7$　**b** $y = 3x + 11$　**c** $y = -2x + 7$

　**d** $y = -2x + 16$　**e** $y = 0.5x + 6$

**8** $x = 1 \pm \dfrac{\sqrt{2}}{2}$

**9** $y = 1.5x + 15$

**10 a** 0.15　**b** 0.6　**c** 0.75

**11** A and B are independent, A and C are not, B and C are not.

**12** A and B are independent, A and C are independent, B and C are not.

**13 a** $\dfrac{1}{15}$　**b** $\dfrac{4}{15}$

**15** Discuss your answer and reasoning with those of others in your class.

**16 a** $p = -0.5, p = 1$　**b** $x = 0, x = \pm\dfrac{2\pi}{3}$

**17 a** 16　**b** 6　**c** −4

　**d** −4　**e** 14　**f** 8

　**g** $x^2 - x + 6$　**h** $4x^2 - 2x + 6$　**i** −1 or 4

　**j** −2 or 7

**18** $a = -0.5, b = 5, c = 0.5, d = 2, e = -3, f = 2, g = 3, h = -3, j = 3, k = 1, m = 11, n = 3, p = -8, q = -20, r = -10.$

**19** $a = 2, b = \dfrac{\pi}{3}$

**20 a** 1.91 rads　**b** 115 cm

**22** $a = 3, b = 2$, C(−6, 0), D(0, −4)

**23** $y = 6 \sin \dfrac{2\pi x}{5} + 2$

**24** $\sqrt{162 - 30\sqrt{2}}$

**25** $x = \dfrac{\pi}{18}, \dfrac{5\pi}{18}, \dfrac{13\pi}{18}, \dfrac{17\pi}{18}.$

**26** $x^6 - 12x^5y + 76x^4y^2 - 192x^3y^3 + 264x^2y^4 - 200xy^5 + 65y^6$

**27** 17%

# INDEX

accuracy, and trigonometric questions xv
addition rule for probabilities 210
adjacent (right-angled triangle) xiii
algebraic equations ix–x
algebraic expressions, manipulating ix
amplitude 158, 159, 163
'and' in probability questions 205
angle of depression xv
  of elevation xv
  of inclination of a line 29
  sum and difference identities 181–3
arc length 33, 36, 43–49
area of a triangle xiv
  given two sides and the angle between
    them 9–12
asymptotes 136, 138

bearings xv
binomial expansion 243

chord length 33, 34
co-domain 53
collinear points 19
combinations 239–42
  and Pascal's triangle 243
complement xvi
complementary events xviii, 219
completing the square 106–7, 120, 124
concavity 88, 138
conditional probability 197–202, 203, 219
coordinates of midpoint 71–2
cosine of an angle, unit circle definition
  7–8, 159
cosine function 159
  positive or negative? 169

cosine ratio xiii–xiv
cosine rule xiv, 16–9, 21–2
counting 235–45
cubic functions 129–134
cyclic quadrilateral 19

degrees, converting to radians 39–40
dependent variable xi
direct proportion (direct variation) viii, 67
discriminant 122
distance between two points 72, 73
domain 53, 54, 55, 139

elements xvi
empirical probability xviii
equally likely outcomes xviii
equation of a circle 148–151
equation of a straight line 64–70, 75–7
exact values (trigonometry) 26–8, 171–3

factorials 237–8
function notation xi
functions xi, 53–4, 56–9
  general 142–7
  graphs of 56, 57
  linear xi, 63–81
  machine analogy xi, 55
  natural domain 55
  polynomial 129–35
  quadratic xii, 85–107
  reciprocal ix, xii–xiii, 136–7
  trigonometric 157–73
  types of xi–xiii

gradient
  line joining two points 70, 72, 73
  parallel lines 80
  perpendicular lines 80–1
  straight line graphs 64–6
graphs
  cubic functions 130
  functions 56, 57
  quadratic functions 86–98
  reciprocal functions xii–xiii, 136–7
  trigonometric functions 157–68

hexagonal numbers 244
horizontal inflection 138
horizontal lines 66
hyperbolas 136
hypotenuse xiii

implied domain of a function 55
independent events 218, 219, 221–3, 226–7
independent variable xi
intersection (sets) xvi
inverse proportion viii–ix

length of the line joining two points 72, 73
line symmetry 139
line of symmetry 86, 88, 91, 94, 96
linear functions xi, 63–81
lines parallel to the axes 66–7